Luis R. Betancourt

FROM THE BIG BANG

TO

CONSCIOUS

INTELLIGENCE

The Great Odyssey of the Universe

To my parents - for my education.

To my wife - for her guidance, love, and loyalty.

To my sons - my reason for pride and improvement.

To my teachers, brothers, and friends - for their wisdom.

Contents

Acknowledgments

I express my deepest gratitude to Mrs. Maria Eugenia Rojas Ohrner C.A.E.1 for the extensive text revision. To Olivier Sarbach Ph.D. for his valuable opinions and recommendations, and to R. Alejandro Betancourt Ph.D. for his advice and the design of most of the maps included in the text. I also like to thank the Ashka Group, from whom I received an enormous amount of information for publication and editing.

Foreword

Dr. Olivier Charles-Albert Sarbach.

*Dr. Sarbach has a degree in Physics
by the Swiss Federal Institute of Technology in Zurich*

Ph.D. in Theoretical Physics from the University of Zurich, Switzerland

He completed four Post doctorates at the following Universities:

*Pennsylvania State University E.U.
(Center for Gravitational Physics and Geometry)*

*Louisiana State University E.U.
(Physics, Astronomy, and Mathematics)*

*California Institute of Technology (CALTECH) E.U.
(Theoretical Astrophysics and Relativity Group)*

*University of California at San Diego E.U.
(Mathematics)*

Since 2007 he has been a full-time "Titular C" Professor and Researcher at the Institute of Physics and Mathematics of the Universidad Michoacana, Morelia, Mexico.

We undoubtedly live in a world where science plays an increasingly important role in our lives. Who can imagine having surgery without anesthesia, traveling from one continent to another without taking a plane, living without your mobile phone and all the accompanying applications? The impressive technological development that we have experienced during the last centuries is mainly due to the curiosity of human beings to know and understand the Universe in which we live, understand the laws of nature, and finally, to know each other itself!

Indeed, what would have happened if our ancestors, when contemplating the sky full of stars at night, had not realized that some luminaries, such as the moon and planets, move related to the stars? What if, in the 16th century, Tycho Brahe had not accurately measured the position of the planets in our solar system? Johannes Kepler could not probably have formulated his laws of planetary motion, and Isaac Newton would not have discovered the laws of mechanics in the seventeenth century. In the same way, observations of electrical and magnetic effects made in the 18th and 19th centuries allowed James Maxwell to formulate the laws of electrodynamics. This theory has significant consequences in our daily lives, such as electromagnetic signals to communicate with each other or fiber optics for high-speed internet access.

Later, the desire to unify the theories of Newton and Maxwell led Albert Einstein to formulate the General Theory of Relativity in 1915, a theory that not only revolutionized our perception of space and time but also led to surprising predictions, like the expansion of our Universe, the existence of black holes and gravitational waves.

These predictions have been experimentally verified in recent years, thanks to a growing wealth of increasingly accurate observations of our Universe.

Among these observations is the recent detection of gravitational waves produced by a binary black hole system by the LIGO-VIRGO collaboration and the first image of the silhouette of a supermassive black hole at the center of the Messier 87 galaxy by the EHT collaboration.

On the other hand, the curiosity to explore the nature of matter on smaller and smaller scales has led to the development of quantum mechanics with unexpected properties such as Heisenberg's uncertainty principle and wave-particle duality. Our understanding of matter's structure is based on the Standard Model that describes fundamental particles and their interactions. This Theory has also been verified with great precision, thanks to experiments like the Large Hadron Collider at CERN.

In parallel to all these advances in understanding the fundamental laws of nature, scientists have analyzed the behavior of complex systems. These systems interact with many atoms or molecules, making it possible to obtain essential achievements in chemistry, biology, medicine, and neuroscience.

We still do not know the full impact these discoveries will have on our future. However, they already had some critical repercussions, such as the transistor that started the development of miniaturized electronics. The Global Positioning System (GPS) must consider the effects of general relativity to locate ourselves with a few meters of precision error. The World Wide Web (www) was invented at CERN, and the development in a record time of vaccines that protect us from diseases, to mention just one couple examples. There is no end in sight. In this century, we can expect other impressive achievements, such as the arrival of humans on the planet Mars, the production of clean energy through nuclear fusion reactors, and the development of artificial intelligence thanks to quantum computers.

On the other hand, significant challenges also await us, such as controlling the problem of global warming and its consequences, eliminating the overexploitation of our planet's natural resources, controlling the spread of epidemics, reducing inequality between social classes and striving to achieve world peace. Therefore, it is increasingly important to guarantee our children a comprehensive education to be prepared for the future and face these challenges. Consequently, it is not only about memorizing knowledge and learning methods to solve problems but also about fostering their understanding methods so that young people can develop novel solutions to the problems that confront us in the future.

The book *"From the Big Bang to Conscious Intelligence: The Great Odyssey of the Universe"* represents an excellent opportunity to embark on a fantastic journey toward the three fundamental questions of origins: the origin of the Universe, the origin of Life, and the origin of the Mind.

Through this journey, Dr. Luis R. Betancourt will take you to know a great variety of absolutely fascinating topics. Among them are the chronology of our Universe, the formation of our solar system and the Earth, the definition and origin of life until the appearance of Homo sapiens, Drake's formula to estimate the number of technologically advanced civilizations in our Universe, the humanity of the future, artificial intelligence, the strange properties of quantum mechanics and their possible applications to understand how our brain and our consciousness, the behavior of animals before earthquakes and the near-death experience. After reading this book, you may decide to participate in the great adventure of science, where there is still much to understand and thus contribute to the significant challenges that await us.

"Never regard study as a duty but as an enviable opportunity to learn to know the liberating influence of beauty in the realm of the spirit for your own personal joy and to the profit of the community to which your later works belong."

—Albert Einstein.

Introduction

The Universe, Life, and Consciousness are the three essential elements of science and philosophy. Its essence determines the existence of the Whole, physical and metaphysical; therefore, understanding its in-depth transcendence poses a significant challenge.

Very close to Christmas 2014, I began to write this book, which had been around in my head for several years. I finally decided to write it from the premise that if you want to read something that has not been reported the way you want, write it yourself.

It was undoubtedly the great Carl Sagan who, with his work and publications, but especially his masterpiece *"Cosmos, a Personal Voyage,"* motivated me. He made me ask myself again about many questions related to the Universe since I was five or six years old. The magic of this great scientist from Cornell University inspired me like thousands of people around the World. Since the mid-eighties, I have searched intensely for information on Cosmology and Astronomy. Finally, a few years ago, Thanks to Dr. Charles S. Cockell from the University of Edinburgh, I delved into the beauty of Astrobiology Science which allowed me to combine my medical training with the space sciences. As a joke, it is said that Astrobiology shares with Theology the characteristic of having more students than objects of study.

As of June 1995, we all had the opportunity to access the images taken by the Hubble Space Telescope and ground-based telescopes worldwide on the Internet through the Astronomy Picture of the Day page. The importance of these images' publication from the beginning was incredible for all those who eagerly awaited every day to see what surprise NASA had in store for us.

The perception of the wonders of the Universe invited us to reflect and ask ourselves deep questions. There are more than 200 million entries with the word *Astronomy* on Google and almost 28 million for *Cosmology.*

Since Carl Sagan's death, many scientists have continued with the noble task of keeping up-to-date information on the advances of astrophysics digested through the mass media. All those who talk about Earth and Space issues always mention the name of the genius[1]. From my point of view, it is genuinely moving that an agnostic man of Sagan's stature died in the intense and deep search for his own God.

A year after I started writing, I came across an article by Jack W. Szostak, an eminent Harvard physician who won the 2009 Nobel Prize in Medicine. In his 2016 article *"On the Origin of Life,"* he mentions in the introduction: « There are three fundamental questions of origins: the origin of the Universe, the origin of life, and the origin of the mind. All of these have been debated for thousands of years and are now the subject of scrupulous scientific research. » Therefore, my task related to these issues had to have depth, updated knowledge, and clarity to publish pleasant and exciting scientific popularization. It was also vital to be impartial in all controversial matters, especially concerning philosophical opinions.

On the other hand, the evolution —since the emergence of the Universe with the formation of galaxies and their planetary systems, the

---

[1] I want to make a special mention to some of Carl Sagan's contemporary and predecessors science popularizers from whom we received outstanding information. I am referring particularly to Isaac Asimov, Desmond Morris, Arthur C. Clark, Stephen Hawking, Paul Davies, Brian Greene, Jacques-Yves Cousteau, Brian Cox, Michio Kaku, and Neil deGrasse Tyson.

appearance of unicellular life, complex life, and finally, the human being– has been a long road full of challenges.

Homo sapiens is possibly a mere accident or evolutionary experiment of nature. However, thanks to this experiment, humanity, with its intelligence and level of consciousness, can ask itself transcendental questions such as whether there is a purpose for the existence of the Universe.

It is said that change is the rule governing everything, and we call evolution successive variations through time. The Universe has evolved, life has evolved, and so has the mind. Finding answers to the most important questions has been the task of humanity for thousands of years, yet we continue in the process without seeing anything more than approximations.

In our world, only a few centuries have been favored by the existence of great geniuses such as Aristotle, Leonardo, Galileo, Newton, Darwin, and Einstein, characters who have given science a strong push with highly significant ideas and knowledge. All the other famous are only bright human beings with high intelligence and creativity, educated and enterprising indeed, but nothing more. Of extreme genius, that rare phenomenon, this mutation that illuminates the path of humanity, how in need we are of its existence. When they are absent, these avatars of knowledge have to be replaced by scientists' groups who add their brilliant minds, each doing their part in generating knowledge. Moreover, we could say that, due to the current level of science and technology, it is challenging to create new knowledge, yet, on the contrary, it happens on a day-to-day basis and at increasing speeds. It is also true that our capacity for wonder decreases at the same rate that computers increase their speed and processing capacity.

Research Centers in all fields of human knowledge, especially in developed countries, receive significant subsidies, as do large Universities that compete with the most important number of research and publications to maintain their international ranking. Nevertheless, knowledge destined to improve life is frequently under the indifference of most governments that are only interested in whether it can be applied for war purposes, population control, or translate into significant economic gains.

For several years there has been a veiled war between brave international groups that want to spread science and culture (UNESCO, Universities, Institutes, even programs such as TED, Coursera, publishing companies as well as internet portals and YouTube) and governments associated with groups of power whose vision includes maintaining its governed population with the lowest possible standards of education that allows them to manipulate their consciences and wills. Few Latin American countries are spared from this terrible, tenacious, and growing action, supported by some developed countries.

However, even in the United States, the level of education has been deliberately lowered. In such a way that, in recent years, postgraduate studies such as doctorates and post-doctorates are occupied mainly by students from other countries, especially the Asian ones, given the low academic level of graduates with bachelor's degrees from most Universities and Technological Institutes in the entire American Union *(Kaku, 2011)*.

In 1981 the *"Reality Club"* was formed, which annually brings together the World's leading intellectuals to address transcendental issues and answer different questions. The club is ruled under the motto: *"Reach the limit of knowledge of the world, seek the most complex and sophisticated minds, put them in a room, and make them ask each other the questions they ask themselves."* Annually they publish a book that answers the best question of the year voted by the group. The meetings and conversations they have annually are genuinely impressive. In 2006, for example, ten celebrities from physics, astrophysics, and cosmology met, of which three were Nobel laureates in physics, and the rest were prominent representatives of CERN, Princeton, Massachusetts Technology, Harvard, Cambridge, and others, to talk about of *"gravity, vacuum, and dark energy."*

In 1996, *"Edge.org"* has appeared as the online version of the "Reality Club" and functions as a documentary archive on the Internet. Edge incorporates all the meetings carried out, with comments and complementary annotations.

The meeting theme in June 2019, *"The Last Unknowns: Deep and Elegant Unanswered Questions about the Universe, the Mind, the Future of Civilization, and the Meaning of Life,"* was published. Twenty renowned specialists covered these topics, including a winner of the Pulitzer Prize and another of the Nobel Prize *(Brockman, 2019)*.

I consider it a duty of all and, to a greater extent, those who have had the fortune to access higher education to spread science and culture to the size of our possibilities. Whether through academic teaching, publishing in different media, or even informally day by day with our family, friends, and colleagues. We must overcome that cancer of society called ignorance that has diminished the overcoming of individuals and nations throughout history. It is the cause of the destruction of the Library of Alexandria, a millennium of the Middle Ages, and the Inquisition's tyranny for almost 600 years. Furthermore, as if that were not enough today, the enormous differences between developed countries and those called the Third World.

I believe we can be satisfied if we have managed to transform and leave our mark on this World at the end of our lives. Not to be praised or remembered but for the merit of having inherited a better world for future generations. It is up to us to make it so.

"There are those that look at things the way they are and ask why?
I dream of things that never were and ask why not?"

—George Bernard Shaw.

The Dawn of
Everything

" If you wish to make an apple pie from scratch,
you must first invent the universe."

—Carl Sagan.

In the first Old Testament book of the Bible and the Torah, Genesis in chapter 1, verses 1 to 3 (*Genesis*, 1: 1-3), written 3,300 to 4,000 years ago, described the World's creation.

1. In the beginning, God created heaven and Earth

2. The Earth was lonely, and chaos and darkness covered the abyss, and the spirit of God hovered over the face of the waters.

3. God said: "Let there be light," and there was light.

In the current scientific conception, the Theory of the Big Bang has explained the origin of the World as follows:

"The Universe originated approximately 13.8 billion years ago with an explosion of unimaginable proportions, arising from a spacetime singularity of infinite density and temperature, zero volume, mathematically paradoxical."

The first version of the beginning of the Universe is part of one of many religious philosophies called *Creationist Cosmogonies*, according to which the Universe and living beings come from specific acts of divine intervention. The second version, called Expansionist Cosmogony, is based on today's most accepted scientific model. It describes how the Cosmos has been expanding since its origin and that at the beginning, it was an infinitesimal point that at a given moment had a titanic explosion, shaping space and time with matter and energy that constitute galaxies, stars, planets, and everything that exists in the physical terrain, and that will continue its expansion until its end in billions of years. Both cosmogonies have a common point: *Omnia Natus est ex nihilo,* "the whole was born from nothing."

Within the creationist doctrines, there are many variants in cultures and regions of the World. However, the stories of most ancient civilizations have in common the participation of a god or gods who gave rise to the earth, the sun, the moon, and stars; also, plants, and animals, and a great majority have narrated the creation of man and woman (anthropogony) as the culmination of this creative work.

There are as many religious or philosophical cosmogonies as probably social cultures the World has had throughout history. The most widespread at present are three:

- The Judeo-Christian God, Yahweh, created everything from nothing in six holy days *(Yahweh, 2015; Jerusalem Bible, 2011; Enns, 2010).*

- Islam reveals the creative power of Allah through three different attributes or modalities: Al-Khaliq as the divine knowledge in which all creation begins; Al-Bari as the divine will through which all creation originates and evolves; and Al-Musawwir as the divine wisdom in which all creation receives its form and reaches its temporal perfection. The Islamic idea of creation is not about understanding it as a beginning and ending event or action but rather the continuing relationship between God and what he creates. The uniqueness of the infinite God and the link to his finite creation is the quintessence of the cosmogony of Islam *(Longhurst, 2013).*

- The Buddhist, whose origin comes from the teachings of Buddha Siddhartha Gautama. Buddhist cosmology is recorded in the Abhidharma scholastic texts from 400 BC. It establishes that a superior being did not create the Cosmos since it has always existed and is part of cycles of destruction and creation; the Universe is born, dies, and is reborn.

Therefore, it is infinite in time and space and has not had a beginning. In addition, in the teachings or sutras of Buddha, the existence of other Universes and different states of being is pointed out *(Landaw, 2011)*. Buddhism and Hinduism[2] are the only religious philosophies in which the time scale resembles modern scientific cosmology. Its cycles range from our common years to Brahma Day, 8.64 billion years long each, and contain even larger timescales *(Sagan, 1985)*.

In Buddhist cosmology, *Kalpas* are called eons or long periods. There are four different types of Kalpas: the common Kalpa lasts about 16 million years, a short-lived Kalpa spans 16 billion years, a medium-size Kalpa is about 320 billion years, and a large Kalpa lasts about 1.28 trillion years. In this philosophical conception, the temporal divisions occur in these enormous proportions because it is established that the Universe is eternal, so they established temporal measures that were probably devised to explain the concept of infinite time. However, unlike this conception, the most prominent time scale towards the past that can be timed according to modern ideas is when "the time begins," the moment in which the Big Bang originates.

However, if we consider the hypothesis of the Multiverse, where infinite and eternal spacetime exists, with universes that continually arise, the religious conception would be very similar to that of the scientific model.

Comparatively to these concepts and seeming somewhat naive, in 1650, the Archbishop of Armagh, James Ussher, in what is now

---

[2] The Hindu cosmogonies derived from Brahmanism are several philosophies very similar to each other and similar to Buddhism, but they have specific differences. For example, these doctrines are theistic, and the origin and existence of the Universe, according to the Upanishad, are derived from three deities or Trimurti: Brahma (the creator god), Vishnu (the maintainer god), and Shiva (the destroyer god). In these philosophies, Brahma's last birth was 156 billion years ago (156 x 1012), and the Universe is created and destroyed every 8.64 billion years, which is a Brahma Day and night equivalent to ~ 1/2 Kalpa of Hinduism. A Brahma year has 360 of its days, equivalent to ~ 3.11 billion years. Each life of Buddha is 100 of his years. After this, he dies to be born again. He is currently 50.1 years old since his last rebirth.

Northern Ireland, wrote the book *"Annales veteris testamenti, a prima mundi origine deducti"* (Annals of the Old Testament, deduced from the first origins of the World). Based on the Bible, he estimated the number of generations, considering the average duration of human life and the central biblical figures from Adam and Eve to the birth of Jesus Christ, and deduced "exact dates" for specific events. For example, the creation of the Earth happened on Saturday, October 22 at 7:00 p.m., 4,004 BC. C., the expulsion of Adam and Eve from Paradise on Monday, November 10, 4,004 BC. C., and the final day of the Universal Flood when Noah's ark rests on Mount Ararat, on Wednesday, May 5, 2,348 BC.

Hindu philosophy's oldest sacred text, the *Rig Veda,* has an agnostic stance on the transcendental question «Who or what created the universe? » *(Kramer, 1986).* The "*Nasadiya-sukta* or Hymn of Creation," the tenth book of the Rig Veda written in the middle of the 2nd millennium B.C., mentions:

> *Who proclaims it? Who knows for sure?*
> *Where did creation come from? Where was he born from?*
> *The gods are after the outcome of this World.*
> *So, who can know its origins?*
> *No one knows where creation came from or whether He did.*
> *He, who contemplates it from the sublime skies,*
> *Only He knows, or maybe He does not.*

Regardless of which cosmological philosophy it is, be it religious or scientific, the essence of each arises from the need to explain our origin and environment; to give meaning to our lives, and provide an answer to the oldest existentialist dilemmas of humanity, such as like who am I and what do I exist for? Where and how did everything come from? These questions and many others closely linked to reality have been and will continue to be one of the most potent engines human beings have in searching for knowledge.

22

FROM THE BIG BANG TO CONSCIOUS INTELLIGENCE

Universe

"The most incomprehensible thing about the Universe is that it is understandable."

—Albert Einstein.

Origin and Evolution

The Universe emerged 13,772 ± 0.039 billion years ago, according to the most recent measurements *(Aiola, 2020)*. The Big Bang, literally the great explosion or burst, constitutes the moment when all matter and energy appear out of nowhere from the point of infinite density and temperature. This theory had its origins mainly in the equations of *"General Relativity"* by Albert Einstein in 1915, which explains the interaction of matter and energy with spacetime in the four dimensions (three spatial and one temporal), adding the gravitational component to the *"Special Theory of Relativity"* that he developed in 1905. Therefore, he also considered it an extension of Newton's Law of Universal Gravitation.

During the development of this Theory, Einstein (1879-1955) realized that his equations did not allow the Universe to remain fixed and invariable in time and space since, due to the force of gravity, they only adjusted to a Universe in contraction. However, the vision of a static Universe was a general idea at that time. Instead of accepting the new result, he decided to trick his equations by introducing the *Cosmological Constant* component (expressed with the mathematical sign lambda "Λ"). Physically, this term translates to a force that causes bodies to reject each other. Einstein adjusted the constant so that this repulsive force exactly balanced gravitation *(Einstein, 1917)*. He thus obtained a static and invariable Universe in time and space. It is said that, years later, he would recognize this as the biggest blunder in his scientific career *(O'Raifeartaigh, 2018)*.

However, a short time later, the works based on general relativity, the Dutch Willem de Sitter in 1917 and the Russian Alexander Friedman in 1922, showed that the Universe is expanding, giving one of the first steps in the construction precisely of the Expansionist Theory. Previously, the American astronomer Vesto Slipher, at the

Lowell Observatory in Arizona, had suggested since 1914 that the Universe could be expanding since he observed through the Doppler effect that more than forty spiral nebulae had large redshifts. That meant they were moving away from our planet at enormous speed *(Slipher, 1915)*.

In 1929 Edwin Hubble elaborated the *Law* that bears his name. Currently, the *Hubble-Lemaître Law* *(IAU, 2018)* indicates that the distance speed between galaxies increases linearly with their distance. «The more distant a galaxy is observed, the redshift is greater. » This speed is determined by a constant with the same name, the *Hubble (Lemaître) Constant,* which is periodically specified; the most recent measurement has a value of $H0 = 67.9 \pm 1.5$ km/sec/megaparsec *(Choi, 2020)*. This law allows us to imagine the Universe as an immense pancake that expands as it is cooked, and all the raisins inside separate from each other. Still, the more distant raisins seem to move away faster than the closest ones to a randomly chosen raisin.[3]

Nevertheless, the amazing original contribution of the astronomer Edwin Hubble occurred six years earlier, on December 30, 1923. He discovered isolated stars located on the periphery of the Andromeda nebula. Due to the luminosity-distance relationship that characterizes these stars of radial pulsation, the Cepheid variables demonstrated that Andromeda was not inside our Galaxy but outside and was a galactic system like ours. These findings were published six years later *(Hubble, 1929)*. This discovery showed that our Galaxy, the Milky Way was not the entire Universe. Therefore, it was made up of several galaxies or nebulae as they were known at the time, proving the hypothesis of the

---

[3] Edwin Hubble scarcely mentioned that his law and the constant arose from many previous contributions. Such as the observations that Slipher had made; to the basis of the astronomer Henrietta Swan Leavitt who throughout her academic life carried out work on the stars called Cepheid variables and developed the metric to calculate stellar distances, also to the results of Knut Lundmark and his method to compare speeds with distances. But mainly thanks to the information he obtained after attending the series of lectures given in 1927 by Georges Lemaître in Holland. Finally, to the unconditional support of his assistant, Milton Humason, at the Hooker telescope on Mount Wilson.

physicist Edmund Halley who had predicted it since the late seventeenth century, and the philosopher Emmanuel Kant, who, in his 1755 work *"The Universal Natural History and Theories of the Heavens,"* developed the hypothesis about the existence of other galaxies external to the Milky Way, calling each one, «Island Universes. » However, the demonstration by Hubble led to acceptance by the scientific environment around the World and the general public *(Lleó, 2015)*.

Another significant contribution was made in 1926 when creating a classification system for galaxies according to their shape, called the *«Hubble Sequence. »* This sequence includes four types of galaxies: *elliptical, lenticular, spiral, and irregular.* This classification is still used today with some later modifications by Vaucouleurs and Sandage and more recent ones.

Meanwhile, in Europe in 1927, Georges Lemaître, a Belgian Jesuit priest and astrophysicist based on the work of the Russian Alexander Friedman, reached mathematical conclusions that gave rise to the so-called dynamic Friedman-Lemaître models, which proved the expansion of the Universe *(Lemaître, 1927)*. These models were ratified in the 1930s by the American physicist Howard Percy Robertson and the Englishman Arthur Geoffrey Walker to produce the Friedman-Lemaître-Robertson-Walker metric. The importance of this FLRW metric is that it represents an exact solution to the complicated field equations of Einstein's general relativity.

However, it is fair to mention that since 1915, the physicist Karl Schwarzschild, through his mathematical work, had already found an exact solution to these equations, a finding that not even Albert Einstein himself had achieved at that time.

His solution was embodied in a letter addressed precisely to Einstein in December 1915; The latter, in turn, thanked and congratulated him shortly after. Unfortunately, Schwarzschild died in May 1916. Years later, his metric would be the basis for understanding some properties of the classic black hole model or *Schwarzschild black hole (Chown, 2019)*.

In 1931, Georges Henri Joseph Édouard Lemaître, supported by his previous works, carried out a deep logical and mathematical analysis of the expansion process of the Universe and postulated in his *article "The Beginning of the World from the Point of View of Quantum Theory"* in the journal *Nature* on May 9, that it originated from the explosion of a "primal atom" understanding by atom as he explained, the indivisible particle of the original Greek conception.

This fundamental fact, supported by his several years of work, gives him international recognition as the *"Father of the Expansionist Theory [4]"* called the Big Bang years later *(Lemaître, 1931)*.

In 1933 *(Fig. 1.1)*, during a conference at the California Institute of Technology, after he finished describing his Theory, Einstein applauded him profusely, commenting: *«This is the most beautiful and satisfying explanation of creation that I have ever heard. »* Furthermore, Pope Pius XII referred to the new Theory of the origin of the Universe as a scientific validation of the Catholic faith; Lemaître, shocked, treated the matter carefully, making him understand how essential it was to separate scientific facts from religious beliefs.

---

[4] In November 2011, Mario Livio, an astrophysicist at the Institute of Space Telescope Sciences in Baltimore, Maryland, published an article in the journal Nature. He recounted his discovery through letters between George Lemaître and William Smart that he had found in the archives of the Royal Society of London. They described the calculations that Lemaître had made about the expansion of the Universe two years before the famous publication by Edwin Hubble (1929) "A relation between distance and radial velocity among extra-galactic nebulae." Lemaître's publication in 1927 (a few months after the letters), where he solved Einstein's equations on the geometry of the Universe and deduced its expansion. It was edited in French and ignored by the scientific community in the United States. This recent article dispelled any doubts and granted George Lemaître solid foundations of the paternity of the Expansionist Theory (Livio, 2011).

Earlier in June 2011, Sidney van den Bergh pointed out that editions of a 1931 English translation of Lemaître's original article appeared to remove various fragments of text that disputed the correlation that came to be known as Hubble's law. The paper also removed Lemaître's version of the Hubble constant from an equation (van den Bergh & Reich, 2011). As we see, history showed us that Edwin Hubble was not such an honest person and took advantage of the work of others to fulfill fame and glory beyond his great achievements.

Figure 1.1   George Lemaître and Albert Einstein in 1933.

Modified from Santa María Madre de Dios Parish. Public domain.

Seventeen years later, in 1948, the Russian physicist George Gamow expanded Lemaître's Theory of the primordial nucleus. The *publications "The evolution of the Universe" and "The origin of the chemical elements,"* together with Alpher and Bethe, confirmed that the Universe originated from a gigantic explosion. In addition, during the first minutes after the explosion, some of today's chemical elements were formed when the

Universe's extremely high temperature and density fused subatomic particles into the chemical elements —hydrogen and helium. This work was continued, among others, by the British astrophysicist Fred Hoyle who enunciated « *the triple-alpha process,* » through which three helium nuclei are transformed into a carbon nucleus. This analysis culminated in the publication *"Synthesis of the Elements in Stars"* in October 1957, in which the synthesis of some other chemical elements of the Periodic Table is described due to their fusion in the nucleus of different types of stars and during supernova explosions, giving this process the name «nucleosynthesis» *(Burbidge, 1957)*. Thus, we now know with certainty how each of the elements in the Periodic Table originated *(Fig. 1.2)*.

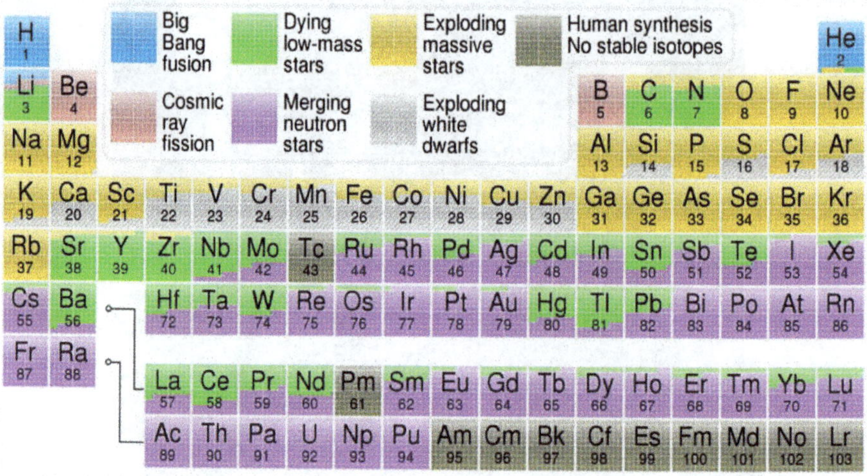

Figure 1.2 *Nucleosynthesis of the elements. Periodic Table.*

Credit: Jennifer Johnson Nov 2017 *Cmglee*. Ohio State University.
Creative Commons Attribution-Share Alike 3.0.

The expansionist theory's confirmation would occur between 1964 and 1965. Arno Penzias and Robert Wilson in New Jersey discovered mysterious microwave radiation that came from something external to their receiver —a Dicke radiometer they used for radio astronomy and satellite communication experiments. Such broadcasts seemed to be everywhere.

This radiation, which they measured with an energy of 2.725 K at a frequency of 160.2 GHz, is the so-called Cosmic Microwave Background (CMB), and it was evidence of the origin of a dense and hot Universe, such as the one proposed by Lemaître and Gamow *(Penzias, 1965)*. Research on this cosmic relic has continued since then, and current data indicate that the background radiation originated 380,000 years after the Big Bang.

By the way, the origin of the name Big Bang is anecdotal and owes its name to Fred Hoyle. Together with Hermann Bondi, Thomas Gold, and others, he developed the *"Steady-State Theory of the Universe,"* in many ways antagonistic to the Expansionist Theory.

On a BBC radio show in London, on March 1949, during an interview, he ironically referred to the alternative theory as ... *"that Big Bang Theory"* and has been known by that name ever since.

In 1916 Albert Einstein edited an article based on his recently published General Theory of Relativity. There, he predicted the existence of cosmic objects that converted part of their mass into energy and released it in the form of ripples that traveled at the speed of light, deforming the fabric of spacetime in its path called gravitational waves. These arose due to the most violent phenomena in the Universe, such as supernova explosions or the collision of black holes.

Sometime later, this prediction began to be the subject of an investigation by various physicists around the World seeking to detect these gravitational waves objectively. Finally, in the mid-1970s, the National Science Foundation (NSF) in the U.S. granted the necessary funds to develop the Laser Interferometer Gravitational-Wave Observatory *(LIGO)*. More than 1,000 astrophysicists from 18 countries currently participate, including the *European VIRGO Collaboration group.*

On February 11, 2016, one hundred years after Einstein's prediction, the NSF held a press conference with LIGO scientists to officially announced that gravitational waves had finally been observed because of the collision of two black holes located 1.3 billion light-years away. It was Einstein's last prediction that was yet to be observed directly.

This finding opened the possibility of using these waves to study the Universe in a new way. Before this discovery, all our information on the Cosmos was through photons at different wavelengths, including radio waves and microwaves. Infrared light, visible light, ultraviolet, X-rays and gamma rays, and the entire energy spectrum of cosmic rays *(Pierre Auger Collaboration, 2020)*. Gravitational waves now allow us to study how black holes are formed and know in detail the life cycle of the Cosmos *(LIGO, 2016)*. In October 2017, Rainer Weiss, Barry Barish, and Kip Thorne received the Nobel Prize in Physics for their decisive contributions to the LIGO detector's development and the objective observation of gravitational waves.

On the other hand, on April 10, 2019, the first photograph of a gigantic black hole of 6.5 billion solar masses was made in the elliptical galaxy *M87,* 53 million light-years from us[5]. This relevant event was carried out using interferometry with a global network of 8 radio telescopes located in Hawaii, Arizona, Mexico, Chile, Spain, and Antarctica, converting the information from the set into an Earth-size radio telescope called the Event Horizon Telescope. *(EHTC, 2019)*. Furthermore, in March 2021 *(Fig. 1.3),* polarized light was added to the image of this black hole, giving new information about its magnetic field *(EHTC, 2021)*.

On May 12, 2022, the International Collaboration of the Event Horizon Telescope, for the first time, released a photograph of Sagittarius A*, confirming the object to contain a black hole at 26,000 light-years from Earth. This image took five years of calculations to process *(Bower, 2022)*.

---

[5] A light-year is the distance that light travels in a vacuum in a year (365.25 days). This distance equals 9.46 trillion kilometers (9.46 x 10^{12}) or 5.88 trillion miles (5.88 x 1012). The exact speed of light in the vacuum is 299,792.458 km/sec. or 186,282.397 mi/sec.

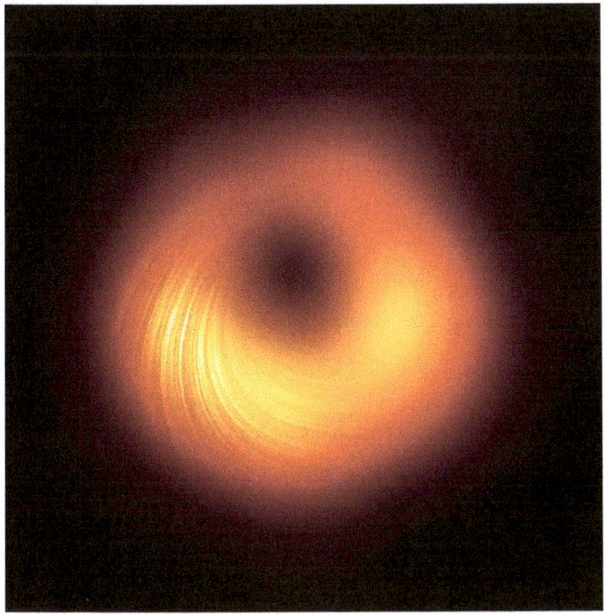

Figure 1.3a   A view of the M87 —supermassive black hole (6.5 billion solar masses) in polarized light.

Modified from The Event Horizon Telescope Collaboration 2021. CC BY 4.0.

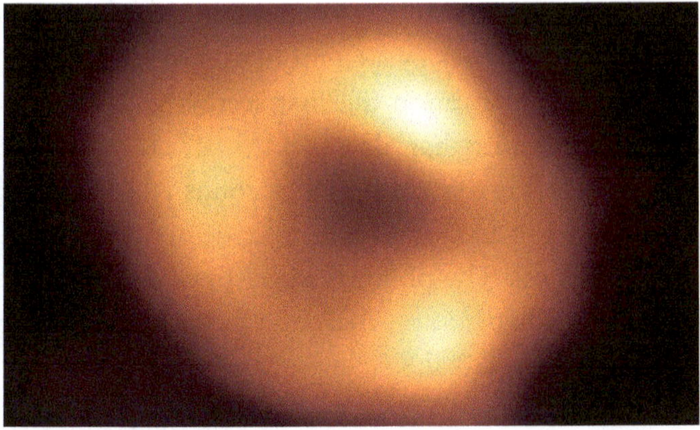

Figure 1.3b   A view of Sagittarius A (4.1 million solar masses).*

Imaged by the Event Horizon Telescope May 2022.

Chronology of the Universe

A sequence of events occurred just after the instant when the Cosmos emerged. The knowledge of the phenomena we are about to outline is the work of several decades of research. The most recent findings regarding the cosmic timeline were published by the European Space Agency (ESA) in 2016 and are the findings of the Planck space telescope mission.

According to the model, time started when the Big Bang occurred 13.799 ± 0.021 billion years ago. The temperature at 10^{-44} sec was 1.4 x 10^{32} K (~100 nonillion Kelvin or "Planck temperature.") —For practical purposes, it is equal to Celsius degrees *(CODATA, 2018)*. Immediately after 10^{-32} seconds, a period of exponential expansion called inflation occurs *(Guth, 1981)*. At that moment, in a micro fraction of a second, the Universe's size multiplies by a factor of 10^{26}, or about 100 septillion times. Light and matter form in the first second and second 100 were coupled; the temperature has dropped to 10^9 K, and dark matter evolves independently. From these first seconds to 380,000 years later, quarks and electrons form atoms (hydrogen, deuterium, tritium, helium, lithium, and beryllium). The photons travel freely and will become the Cosmic Background Radiation (CMB); the temperature throughout the Cosmos has dropped to 3,000 Kelvin.

The period that occurs from this moment to 200 million years later is collectively called the "dark ages" *(Hatch, 2020)*, where the gravity of dark matter exerts its influence on atoms, and the absence of light is the main characteristic of this stage *(Fig. 1.4)*.

Two hundred million years after the Big Bang formed the first stars, and another 180 million years later, galaxies (quasars) *(Vink, 2020)*. Slowly the Cosmos begins to fill with light again; the photons break and separate the atoms, reionizing the Universe.

Finally, five hundred million years after the Big Bang, the first clusters of galaxies are integrated. Our Milky Way was born approximately 13.2 billion years ago, and the solar system 4.568 billion years ago, in the arm or spur of Orion, half the distance between the center and the galactic periphery *(Planck Collaboration, 2016; Connely, 2017).*

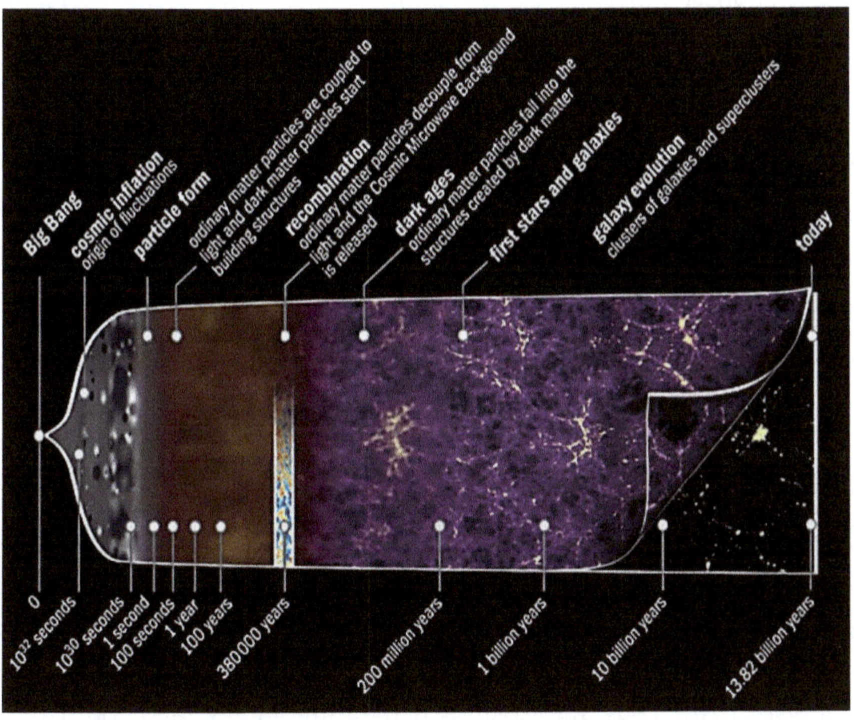

Figure 1.4 Chronology of the Universe.
Credit: ESA / Planck Mission / NASA. Public domain.

Structure of the Universe

Now, we must fully understand how the Universe is, at least as far as the most recent knowledge allows us. The Universe is everything that exists; it is all spacetime and its matter and energy content. Space-time has no borders; therefore, it is unlimited. The word Cosmos refers to the orderly and harmonious Universe as opposed to the word Chaos. Nobody knows its dimensions; we can only perceive a part of it called the observable Universe, and it has a diameter of 93.2 billion light-years where approximately 2 trillion galaxies reside ($2.0 \pm 0.6 \times 10^{12}$) with ~ 600 quadrillion stars (6.0×10^{17}) *(Conselice, 2016)*. This figure is ten times higher than the calculation done 11 years earlier *(Gott, 2005)*.

The Universe, observed with the naked eye, is not very uniform; hence the boreal and southern constellations have been imagined to be drawn, but at large scales (distances greater than 100 megaparsecs or ~ 326 million light-years), the Universe is very homogeneous and isotropic, spatially flat (Euclidean), four-dimensional (space-time), and congruent with General Relativity.

A homogeneous and isotropic Universe means that mass and radiation —due to gravity— are distributed with the same average density everywhere (homogeneous) and in all directions (isotropic) *(Fajman, 2020)*. Thus, the Cosmos is the same in any direction for any observer, and there are no privileged observers since there is no center. The fact that the Universe has homogeneity and isotropy on a large scale is part of what is known as the *Cosmological Principle*. From this, it follows that all the laws of physics are the same anywhere in the Universe.

In recent years it has been possible to discover structures formed by massive clusters of galaxies with dimensions of several billion light-years that do not appear in other areas of the visible Universe, as well as areas colder than others. So homogeneity and isotropy are less evident in various regions of the Cosmos; therefore, according to general relativity, both properties are imperfect *(Ashtekar, 2021)*. Otherwise, the Universe would be a kind of soup of matter and energy perfectly distributed as it was precisely at the moment of its appearance, and galaxies or any other cosmic structure would not exist *(Pettinari, 2015)*.

In addition to the essential characteristics that make up the Cosmological Principle, we must know another crucial aspect of the Universe's shape *(Muñoz, 2003)*. The basis for this shape lies in the local geometry of the observable Universe. There is a density parameter called Omega (Ω) related to the curvature of space so that the Universe can be flat, closed, or open according to the values of this parameter.

Omega is equal to the average density of the Universe formed by the mass and energy it contains, divided by the density of the critical energy that depends on the expansion speed and the universal gravitation constant.

If the average density and the critical energy are equal, then: $\Omega = 1$ that is, the curvature of the Universe is zero; therefore, the Universe is flat. If the curvature is positive $\Omega > 1$ because the average density is more significant than critical energy, then the Universe has spherical geometry and is a closed universe *(Friedman, 1922)*. Finally, if the curvature is negative $\Omega < 1$, that is, the average density is less than the critical energy, then it is a universe of hyperbolic or saddle geometry, equivalent to an open universe *(Friedman, 1924)*.

Currently, the combined measurements of the curvature of the space made by the Wilkinson Microwave Anisotropy Probe *(WMAP)*, the Sloan Digital Sky Survey that measures baryonic acoustic oscillations, and more recently by the Planck space telescope, show that $\Omega = 0.9992 \pm 0.002$, which is coincident with a flat Universe and congruent with general relativity *(Fig. 1.5.1)*. Thus, a cosmos of Euclidean or plane spatial geometry is a universe that will continue to expand endlessly until its energy is entirely exhausted. This ending has been called Big *Freeze (Bhattacharya, 2019)*.

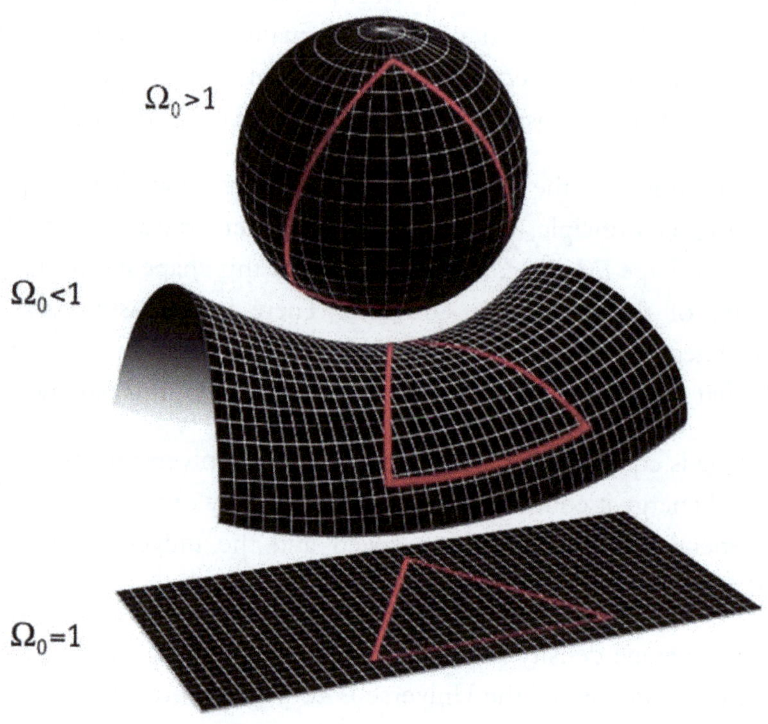

Figure 1.5.1   Three possible geometries of the Universe (Friedman, 1922 & 1924).

Credit: NASA Gary Hinshaw. Public domain.

Then, since this expansion is increasingly accelerated due to the energy of the vacuum or dark energy that we will talk about next, it has been theorized that a rupture will occur in the space-matter-energy structure of the entire Universe so that this end of its existence has been called the *Big Rip*.

Finally, regarding the knowledge we have of the matter-energy composition of the Cosmos, it is the result of WMAP measurements since mid-2001. More recently, through the results of the observations of the Space Telescope of the European Planck mission, of the BICEP 2 and 3 (Background Imaging of Cosmic Extragalactic Polarization) radio telescopes located in Antarctica, the Keck Observatory in Hawaii, from 2009 to 2016, and particularly the Department of Physics and Astronomy from the University of California at Riverside from September 2020. These measurements have a margin of error of less than 0.2%. The results of these indicate that the Universe is constituted as follows: Dark energy 68.5%, Dark matter 26.6%, and Atoms 4.9%, of which hydrogen and free helium is 4%, stars 0.5%, neutrinos 0.3%, and heavy elements: 0.03%, photons 0.01% and black holes 0.005% *(Planck Collaboration, 2016; Abdullah, 2020)*. Unfortunately, the amount of antimatter in the Universe is unknown, so it is impossible to include a percentage in the composition of the Universe.

The incredible thing about these results is that we can only observe less than 5% of all the matter and energy that make up the Universe, even with the most advanced technology. This fact is equivalent to a baryonic matter density of just one proton for every four cubic meters of space volume *(NASA / WMAP Science T, 2014)*.

We know about the existence of Dark Energy because observations of supernova explosions in various galaxies show that something is acting as a repulsive force that accelerates the Universe in all directions, opposite to the force of gravity. This repulsive force began at the moment of the Big Bang, as did the force of gravity, but the latter was the one that maintained control over the expansion of the entire Universe for 7.642 billion years.

Approximately 6.145 billion years ago, the expansion of the entire Cosmos began to accelerate progressively because the gravitational force between galaxies decreased as they moved away from each other *(Alriga, 2020)*. Since then, this vacuum or dark energy capable of creating more space in the entire Universe has dominated over the gravitational force.

Until a few years ago, the intuition was precisely due to the force of gravity between galaxies, clusters, and superclusters of galaxies; after so many billions of years, the expansion of the Universe should be slowing down. However, the exact opposite is happening; it is accelerating. This discovery occurred in 1998, based on observations about the degree of luminosity of type 1A supernovae concerning their distance and the presumed recession speed, which was much higher than calculated. For this reason, in 2011, Saul Perlmutter, Brian Schmidt, and Adam Riess received the Nobel Prize in Physics. Their observations were confirmed in 2017 by David Cinabro and his team from the Universities of Wayne and Chicago *(Cinabro, 2017)*.

To mathematically explain this invisible expansive force, it required to bring back to life "the biggest blunder in Einstein's scientific trajectory," the *Cosmological Constant,* which is equivalent to the energy of a vacuum and scalar fields, which seem to explain the expansive behavior of the Cosmos. One explanation for dark energy is that it is a property of space. Albert Einstein was also the first to realize that « empty space does not mean that it is equal to nothing, » The observations of hundreds of galaxy clusters with distances between 0.76 and 8.7 billion light-years show that the amount of dark energy has not changed in billions of years. It also supports the idea that dark energy can be explained through Einstein's "Cosmological Constant," which is equivalent to the energy of empty space *(Morandi, 2016)*.

Recently it has been inferred that the nature of dark energy has been stable since the beginning of the Universe probably means that it is a product of the Multiverse, but this is not easy to *verify (Durrer, 2011; Huterer, 2018)*. On the other hand, the existence of Dark Matter represents almost 82% of the matter in the Universe, and its presence is deduced based on its gravitational effects on visible matter and radiation in galaxies and groups of galaxies.

We will explain this in simple terms: if we observe how the planets in the Solar System behave, we know well that, due to the balance between their inertia and gravity (angular momentum), the closer they are to the Sun, the greater the speed at which they perform an orbit. For example, Mercury takes 88 days for a complete revolution at ~ 48 km/sec, Earth 365.25 days, or an Earth year at ~ 30 km/sec, while Neptune takes almost 165 years at ~ 5.4 km/sec *(NRAO, 2016)*.

The Milky Way[6], our Galaxy, is a barred spiral one. It has a rotational movement with its center as its axis. The population of stars is probably more than 200 billion, and they are distributed in its arms, with a greater concentration as we advance to the galactic center. Therefore, we could assume that, like planetary kinetics, stars farther from the center would take much longer to complete a full orbit (at the galactic position of the Sun, the orbital time is ~237 million years). However, this is not so; the Galaxy rotates similarly to a disk with slight delays in the periphery. If we transfer to our solar system, the rotation time range of the Galaxy is ~220 million years in the area adjacent to the center and ~360 million years on average in the periphery *(Gerhard, 2008)*; this It would then mean that Neptune would orbit the Sun in just 144 days with Mercury as its reference *(Fig. 1.5.2)*.

The mass required to achieve this impressive gravitational pull is vastly more significant than the amount of observable matter. So, as it seems, there is something more there than stars and planets. Since it does not emit or absorb light or any other electromagnetic radiation and is only detectable by its gravitational effects, it has been called *"Dark matter."*

In addition to its effects through the rotation of galaxies, its existence has also been proven through the distortion of light when passing through galaxy clusters and in simulations of the early Universe

---

[6] The Milky Way does not have a flat disk shape as previously thought but is deformed and twisted like a potato chip " ~ " *(Skowron, 2019)*; Its radius is approximately 95,000 light-years *(López, 2018)*, and it weighs 1.5 billion solar masses (1.5×10^{12}), 85% of which is dark matter *(Watkins, 2019)*. ~

that require the presence of dark matter for the formation of the galaxies. In fact, since 2008, the Fermi gamma-ray space telescope has detected enormous amounts of dark matter in and around dwarf spheroidal galaxies.

The dark matter hypothesis has a long history. Since the late 19th century Lord Kelvin predicted the existence of dark bodies in the Milky Way.

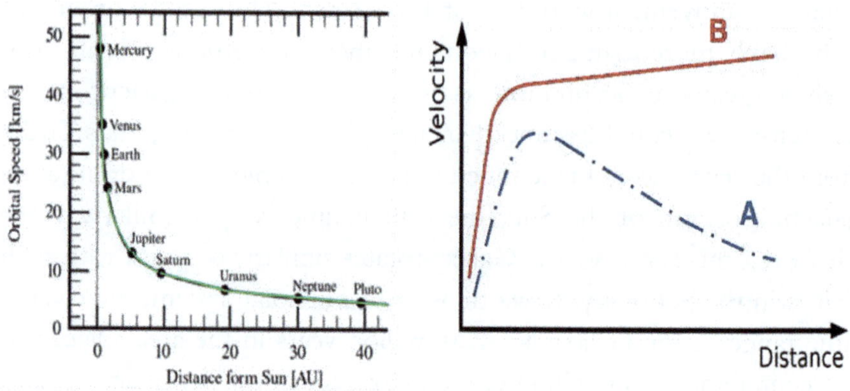

Figure 1.5.2   Left - Orbital speed of the planets concerning their distance from the Sun. Right   -   Galactic orbital rotation speed: expected (A) and observed (B). Note how the expected galactic curve is similar to the planetary one.

Modified from: Torres-Silva (2008) A Metric for a Chiral Potential Field *Rev Chil Ing* 16 (sp. number):91-98, and Phil Hibbs GalacticRotation2. Inkscape. Public domain.

In 1906, Henri Poincaré first used the French term "matière obscure" when referring to Kelvin's observations. Subsequently, the contributions of Kapteyn (1922), Lundmark (1930), Oort (1932), and Zwicky (1933) were added. In addition, one hypothetical particle that constitutes dark matter has been called "Weakly Interactive Massive Particle" or WIMP. Other hypothetical particles that form in the stars' nucleus are axions or axion-like particles that can become gamma rays or X-rays and return to the state of particles *(Fraser, 2014)*. Finally, a part of the matter that today is considered dark can be baryonic matter;

this fraction named "massive astrophysical compact halo object," known by the acronym MACHO (Massive astrophysical compact halo object), which includes black holes, quark, and neutron stars, white dwarfs, brown dwarfs, and isolated planets, all of them due to their characteristics are complicated to detect individually, especially those located in the halo that surrounds all galaxies.

In conclusion, the *"Standard Model of Cosmology"* tells us that our Universe is ~ isotropic and homogeneous, has Euclidean spatial geometry, and therefore is flat. Since its structure is mainly made of energy and dark matter, it is congruent with the Lambda-CDM or ΛCDM (Cosmological Constant - Cold Dark Matter) model, derived from the Friedman-Lemaître-Robertson-Walker metric of general relativity. The model represents a theory of concordance with the Big Bang because it explains the Cosmic Microwave Background electromagnetic radiation (CMB) observations and the large-scale structure, acceleration, and expansion of the Universe. It is the simplest known model that agrees with all the most recent observations. One of its predictions of the end of our Cosmos, in addition to the *Big Freeze* (in trillions of years), is the *Big Rip* that could occur in about 15 to 22 billion years. This Big Rip is not yet fully understood, as it depends precisely on dark energy, whose knowledge is still evolving.

Similar but more intense energy could cause the inflation period at the beginning of the Big Bang. However, it only persisted for a fraction of a second, in such a way that some current theories explain that there may be more than one repellent energy or the same energy with various forms of behavior *(Caldwell, 2003; Hughes, 2019)*. These have multiple names, such as Vacuum energy, Quintessence, and Phantom energy, each with its mathematical basis. One of them, the "quintessence," explains that when dark energy collides with baryonic particles, it can become dark matter; however, this is only a hypothesis.

Thus, the current scientific status concerning the composition and final destiny of the Universe is very similar to what happened a century ago when the first hypotheses regarding its origin began.

The Microcosm and the Standard Model of Particles

Now let us talk a little about the constitution of matter and energy. In recent years, a large amount of evidence, such as the discovery of the Higgs Boson —which gives mass to elementary particles— at the European Center for Nuclear Research (CERN) *(Conseil Européen pour la Recherche Nucléaire)*, has come to strengthen the "Model Standard of Particles," one of the bases of modern Physics that explains how the basic building blocks of matter interact and organize through the four fundamental forces that are: the *strong nuclear force*, the *weak nuclear force*, the *electromagnetic* and the *force of gravity (Charitos, 2014)*.

Quantum Mechanics explains three fundamental forces: the strong nuclear force, the weak nuclear force, and the electromagnetic force. On the other hand, the Theory of General Relativity explains gravity and defines it as the deformation of spacetime. This fundamental force is the weakest despite affecting the largest structures and the greatest distances. For example, if given a value of 1 for force relative to gravity, the strong nuclear force would be 10^{38}, the electromagnetic force would be 10^{36}, and the weak nuclear force would be 10^{25}.

Dr. Gordon Kane, professor at the University of Michigan and leader in theoretical and phenomenological particle physics, defines the Standard Model as follows: *«It is a comprehensive theory that identifies basic particles and specifies how they interact. Everything that happens in our Universe (except the effects of gravity) results from the particles interacting according to their rules and equations. »*

The model shows that two fundamental particles exist in nature: -*Fermions and Bosons- (Fig. 1.6)*. Fermions are matter's primary or elemental constituents, while bosons are the mediators of the fundamental forces that govern the particles *(Kane, 2003)*. Leptons and hadrons integrate *Fermions*. Quarks form hadrons, so the fundamental or elementary particles of *Fermions* are leptons and quarks because it has not been proven with certainty that they have an internal structure made up of even smaller particles.

There are six *leptons* (electron, muon, tau, electron neutrino, muon neutrino, and tauon neutrino) and six *quarks* [up quark (u), down quark (d), charm quark (c), strange quark (s), bottom quark (b) and quark top (t)]. The quarks also have a property called color *(quantum chromodynamics)*, and each has three colors (red, green, and blue); therefore, there are 18 quarks. Nevertheless, there are 12 leptons and 36 quarks since each particle corresponds to an antiparticle.

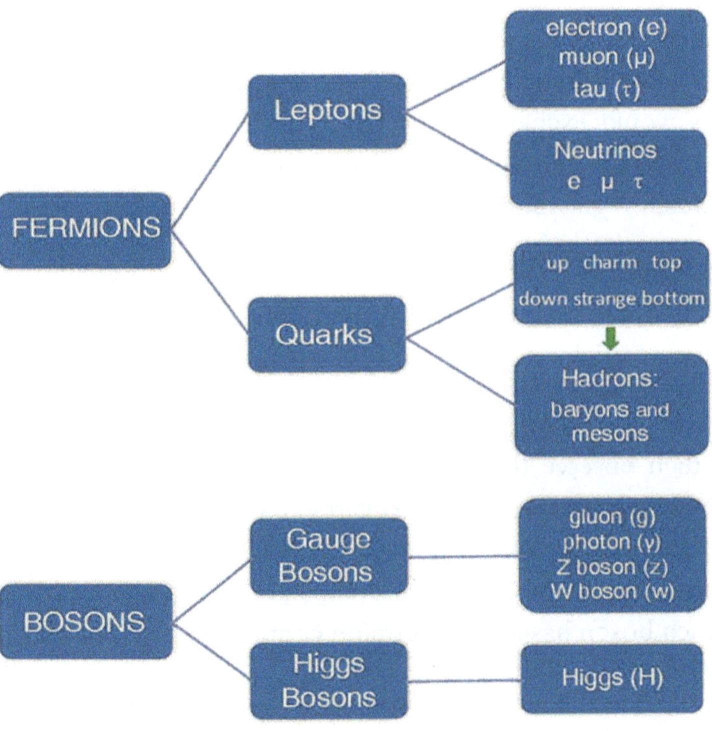

Figure 1.6 Scheme of the particles and elemental forces.
Credit: Own original work.

The other Fermions in the model are *hadrons*, particles with an internal structure because they are, in turn, made up of quarks.

There are two types of hadrons: baryons (the best known are protons and neutrons), made up of three quarks or three antiquarks, and mesons made up of a quark and an antiquark (for example, pions and kaons), although it has been discovered exotic hadrons consisting of 4 and up to 5 quarks *(Aij, 2015)*. Currently, ~260 hadrons, "120 baryons + 140 mesons," have been detected *(Moreira, 2009; Tanabashi, 2018)*.

On the other hand, the particles' fundamental interactions or exchanges have to occur through the second basic type of particles in nature. The energy carriers are the *Bosons*, constituted by gluons in the strong nuclear force, the W $\pm$ and Z bosons in the weak nuclear force, photons in the electromagnetic force, and gravitons (not yet discovered) in the gravitational interaction. That is, charged-color particles (quarks) would interact by exchanging gluons, weakly charged particles would exchange W +, W-, and Z particles, and electrically charged particles would interact by exchanging photons. In contrast, mass particles (gravitines) would exchange gravitons.

The Higgs boson (or Higgs bosons, because it is hypothesized that more could be found in the future) gives rise to the mass of all elementary particles, including the W (weak) and Z (zero) bosons *(Fig. 1.7)*.

We then observe that, although there is a correlation between the particles of matter and the energy carriers, the so-called supersymmetry extends the number of particles of the Standard Model. Each particle corresponds to a supersymmetric companion called a super companion. Thus, each boson has a super fermion companion and vice versa.

The super companions of fermions are bosons and have names that begin with the letter "*s.*" Thus, the electron has the selectron as a super companion and the quarks, the squarks. The super companions of bosons are fermions with names that end in "*ino,*" so that of the photon is the *photino*. For the graviton, the *gravitino*, that of the gluon, *glüino*; that of the W boson, *wino,* and that of the Z boson, *zino,* and this relationship has been called supersymmetry. However, despite initial theoretical arguments from the 70s and 80s, no experimental evidence has been found that absolute supersymmetry exists in nature *(CMS Collaboration, 2021).*

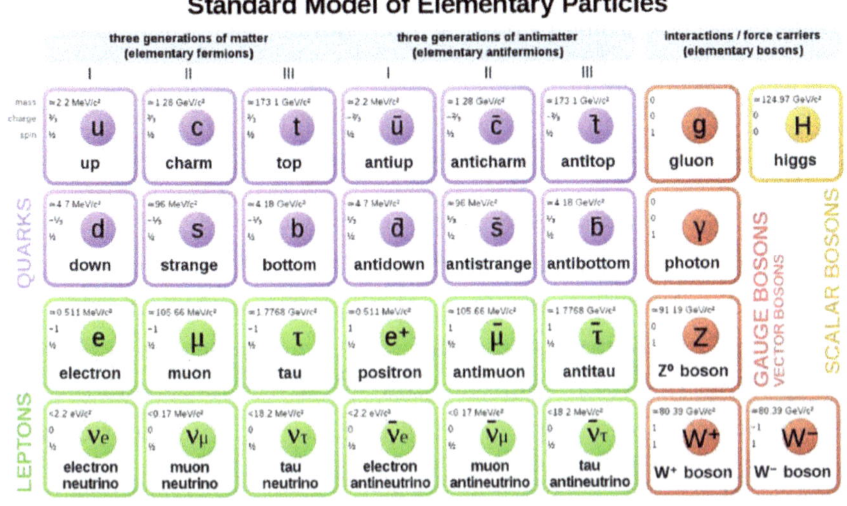

Fig 1.7 Standard Model of Elementary Particles. Matter and antimatter.

Credit: *Cush's* own work (October 14, 2018). Public domain

Dr. Gordon Kane, with whom we have already spoken, relates that at a conference in South Korea in 2002, Dr. Stephen Hawking approached him and proposed a $100 bet that the Higgs Boson would not be found. Ten years later, on July 4, 2012, a check for that amount was sent to his home; given that, during the 36th International Conference on High Energy Physics in Melbourne, Australia, its discovery was officially announced by the directors of CERN. *(Aad et al., 2012).*

The exposition of current knowledge of the microcosm would not be complete without mentioning String Theory, developed between the

60s and 70s and expanded in the following two decades. The essential concept of string theory is this: all the fundamental forces of the Standard Model are different manifestations of a primary object: "a string." The Theory of Superstrings also includes fermions and not only bosons, so it includes elementary particles and not exclusively fundamental forces.

The size of the strings is infinitesimal, much smaller than the leptons or quarks. So, according to the way they vibrate, they will manifest as a photon, an electron, a neutrino, or a quark just, as it happens when a string instrument is played; different vibrations produce different musical notes.

The strings have the Planck length, equivalent to 1.6×10^{-35} m, while the size of the electrons and quarks is $\sim 1.0 \times 10^{-18}$ m. There are five string theories, and altogether they are called Theory M; within these is the Theory of "Branes or membranes." The Branes (p and d) are dynamic objects that can propagate through spacetime according to the rules of quantum mechanics, and each brane or membrane corresponds to a Universe.

The strings move in a spacetime of more than four dimensions: 10, 11, and even 26 dimensions *(Zwiebach, 2009)*. Thus, in superstring theory, 10,500 possible universes result from explaining the seemingly arbitrary values of the Universe, such as the Planck length or the mass and charge of the electron.

String Theories have, from their inception, been the subject of controversy between staunch defenders and bitter opponents. The real value of these results is that they are a physical model that tries to unify the four forces of nature. Many physicists consider that String Theory is not physics but mathematics. There are even those who think that it is not science because the string theory, to be able to explain in a unified way everything we already know, has to resort to a degree of abstraction so high and mathematical so complex that most of the calculations do not have any physical meaning.

And not only that, it is impossible to carry out an experiment proving this model. However, the theory continues to evolve, and that is because it has defenders of the stature of Juan Maldacena, Michio Kaku, Joseph Polchinski, Edward Witten, Dieter Lüst, and especially Brian Greene, a mathematical genius and theoretical physicist from the University of Harvard, Cornell, and Columbia. Moreover, he manages to keep the interest alive with popular science works for the general public, such as *"The Elegant Universe,"* and offering lectures at the *World Festival of Science*, of which he is a founder, and at *TED* (Technology, Entertainment, and Design).

String Theories in the coming years should prove their value as a scientific model by coherently explaining the Universe[7].

On the other hand, the Quantum Field Theory has recently gained more and more acceptance. This theory indicates that individual particles are only a manifestation of their associated quantum fields. Moreover, these probabilistic fields that fill the entire Universe have a non-linear regime that manifests as waves and particles *(Crespo, 2020)*.

Initially, Quantum Physics dealt only with quantifying particle motion, setting aside the classical electromagnetic field, hence the name *quantum mechanics*. Later, the electromagnetic field was also quantified. Finally, even the particles themselves were re-represented through "quantized fields," which resulted in the development of —*quantum electrodynamics* and *quantum field theory*. Thus, by convention, the original form of particle quantum mechanics is known as the first quantization, while quantum field theory is the second quantization *(Espichán, 2017)*.

Now then, we know that the intimate structure of the Cosmos is organized hierarchically. Thus, matter and energy start from simple subatomic particles to gigantic galactic superclusters; subatomic

---

[7] There is methodological reasoning called the Occam's razor (attributed to the Franciscan friar and philosopher William of Ockham), which says that "entities should not be multiplied without necessity," inaccurately paraphrased as "the simplest explanation is usually the best one," consequently, this reasoning does not favor the M-Theory.

particles group together to form atoms, forming molecules and integrating inorganic and organic polymers. Organic polymers will form cells, and later, tissues, organs, and simple and complex multicellular organisms that, together with unicellular beings, will integrate biospheres

From the cosmic dust by the effect of gravity, new stars will be born that will form planetary systems; billions of them will constitute galaxies, groups of galaxies will form clusters, superclusters, and finally, the entire Universe. *Table 1.1* shows the conformation of physical, chemical, and biological hierarchies.

The lower and central part of the Table mentions the *"Loop Quantum Gravity" (LQG)*, a theory initially developed by the Indian theoretical physicist Abhay Ashtekar in 1986 and supplemented in the 90s by Lewandowski, Rovelli, Smolin, Jacobson, Gambini, Pullin and Thiemann. This theory arose when it was realized that Einstein's equations could be written similarly to the Yang-Mills theory, which allows the unification of electromagnetic, weak, and strong forces.

These concepts helped a lot because although it is not yet known how to quantize gravity directly through the Einstein equations of General Relativity, it is known how to do it with the Yang-Mills theory, which led to the intersection of graphs called loops or curls.

To understand LQG, we must remember that energy does not propagate continuously but in discrete "quanta" or packets of energy. In the LQG theory, the same happens for spacetime that cannot be observed as a continuum and, therefore, cannot be infinitely divided, but also can be quantized in specific units of area and volume *(Smolin, 2004),* like the voxels[8] of 3-D television, and the distance these spacetime quanta has is the Planck length (1.6 x 10^{-35} meters).

---

[8] A voxel is a volumetric pixel. A pixel is the smallest homogeneous color unit part of a digital image with two dimensions (square). Instead, a voxel is a cube, and therefore together, they originate 3-D digital images on screens.

Complex entities are built from elementary constituents

COMPLEXITY	UNIFIER	ELEMENTS
Millions of proteins, carbohydrates and fats	Polymerization	22 amino acids form all proteins ~12 monosaccharides form most carbohydrates ~25 fatty acids form most fats
Trillions of unique, individual living things; billions of species	DNA	4 nucleotides bases on a double helix chain
Billions of chemical compounds	Periodic Table	90 naturally occurring chemical Elements (Element families -metals, inert gases, etc.- hint at deeper layer of organization)
Thousands of ions and isotopes	Atomic and nuclear models	3 subatomic particles: electron, neutron, proton (Discovery of new "elementary" particles motivated the search for a model that could accommodate them all)
Hundreds of subatomic particles	Standard model	6 quarks, 6 leptons, 4 bosons – electric charge, mass, color charge, spin and flavor (always recognized as incomplete since it is missing gravitation, dark matter, dark energy)
All standard model particles plus particles yet to be discovered	Superstrings? M theory? Loop Quantum Gravity?	1 string in 10 dimensions with multiple modes of vibration? space and time are constructed from elementary entities? (the theory of everything?)

Table 1.1   Hierarchies of physical, chemical, and biological components.

Modified from Glenn Elert (2021). Beyond the Standard Model. The Physics Hypertextbook.

Thus, the spacetime of the LQG is structured in networks of loops connected and are called spin networks that are studied mathematically through the Graph Theory that calculates the possible "nodes" and their distribution to constitute the network. A spin-lattice represents the dynamic quantum state of a gravitational field; thus, the deformation of spacetime due to gravity originates precisely here, in quantum lattices. *"A spin network is not in time and space, but rather it is time and space,"* which is the core of this theory *(Moreno, 2018)*. LQG fused quantum mechanics and general relativity, and then gravity is formulated identical to the rest of the interactions of the standard model of particle physics. Consequently, this theory rivals the Theory of Superstrings since it also tries to unify the four interactions or forces of the Universe to constitute the *Theory of Everything.*

It has been thought that both theories could be complementary because the Superstring theory emerged from particle physics and was initially formulated as a theory that depended on a fixed spacetime, either flat or curved, and obeyed the equations of Einstein. But, on the contrary, Loop Quantum Gravity was formulated independently of this fixed spacetime *(Gambini, 2020)*. Thus, some theoretical physicists believe that LQG and superstring theory may be two aspects of the same underlying Theory and whose synthesis will lead to a complete theory of quantum gravity. Still, now, it is just speculation with little evidence.

On the other hand, one of the major attractions of the LQG is that it avoids the zeros and infinities of the singularity present in the Theory of General Relativity when applied to the Big Bang and the Big Crunch [9] *(Struyve, 2017)*. The Theorem of Stephen Hawking and Roger Penrose *(Hawking, 1970)* used relativity, demonstrated the existence of singularities within the General Theory of Relativity, and constructed one of the best-known definitions of the Big Bang. *"An explosion of*

---

[9] The Big Crunch is a hypothetical scenario for the final fate of the cosmos. The expansion of a closed universe eventually reverses and collapses again, leading to the cosmic scale factor at the last instant reaching zero, an event potentially followed by a universe that begins with another Big Bang.

unimaginable proportions, arising from a spacetime singularity of infinite density and temperature, zero volume, mathematically paradoxical."

However, the Big Bang does not refer to an "explosion" in an existing space. Still, it designates the expansion and joint creation of matter, space, and time from what is known as a singularity, a geometric point without any dimension. Furthermore, imagine the Universe in reverse time, where it gets smaller and smaller, but the amount of matter is the same. Thus, the density of matter and energy increases until it reaches the point where it becomes infinite, higher than the maximum possible physical density, which is the Planck density (5.15×10^{96} kg /m^3). This figure is equivalent to 10^{23} solar masses compressed in the space of a single atomic nucleus and is the limit for the LQG; above this limit, the equations fail because mathematically, it is impossible to treat infinite quantities.

That is why the theories about the Big Bang do not describe the event itself but the evolution of the early Universe immediately after the event in a time range that spans from the Planck time (5.39×10^{-44} seconds after the Big Bang) until approximately 380,000 years later *(CODATA, 2018).*

As in the LQG theory, there are no zeros or infinities; by quantifying the time and limiting its values, the energy retention capacity of these loops is also limited until they are saturated as they approach values close to the singularity. The result is a force of repulsion. Energy cannot increase to infinity; quantized (discrete) time prevents it. The LQG theory predicts that no such Big Bang singularity exists because a force opposes it when trying to compress the Universe infinitely. Therefore, a minimum size can reach from where it is generated by a counter expansion force, which explains the current kinetics of the Universe, thus concluding that before the Big Bang, there was another contracting Universe.

This kinetics is the so-called *Big Bounce* that constitutes the cyclical model or oscillating Universe. One interpretation of the Big Bang results from the collapse of a previous Universe. Probably, the Cosmos that preceded the actual Universe, according to this Theory, had an inverted spacetime, among other characteristics. We could imagine and ask countless questions such as whether it had the necessary properties to have hosted life and if, through this Theory, we could conclude that the Universe is eternal, without beginning or end, as it has had an infinite number of rebounds[10] *(Santaolaya, 2020)*. Despite trying to be a Theory of Everything, the LQG theory is still incomplete because physicists have no consensus on quantifying gravity in general relativity.

We will close this section by mentioning that there are still considerable gaps in Astrophysics and Particle Physics that must be filled with new and essential knowledge, such as the definitive Theory of Everything. However, this will surely be achieved since Einstein once said: « Although the Universe is immeasurable, it is comprehensible, » and the English physicist Freeman Dyson argued, « *the laws of Nature seem to have been designed in a simple way to make the Universe as interesting as possible.* »

After reviewing the Universe's beginning, its evolution with the formation of stars, planets, and galaxies, and its constitution through the Standard Theory of Particles and others, we will now know how the Earth originated.

---

[10] In the "singularity," space-time is infinitely curved, space has zero volume, and the time comes to a complete standstill. In a space with zero volume, the matter-energy density becomes infinite, which causes the temperature to rise to infinite levels. At the precise moment when the Big Bang occurs, the singularity disappears. The time can be measured from its minimum expression, Planck's time, equivalent to 10-44 seconds. Space ceases to have zero volume, and now the entire Universe has the Planck length of 10-35 meters. Infinite density and temperature cease to exist, instantly giving rise to a very high but measurable matter-energy density and temperature. Therefore finite.

Birth and Evolution of the Earth

Chronologically, the age of the Universe is three times that of the Earth. Our planet was formed ~ 4.567 billion years ago by the accretion of "planetesimals" in the inner part of the protoplanetary disk at the center of which was a newly born third-generation yellow dwarf star that we call the Sun.

Since its formation, the history of events on Earth has been carefully arranged in time within the *International Chrono-stratigraphic Table*. This International Organism divides time in a complex way into geochronological equivalents that include a Supereon of ~ 4.026 billion years, four eons each between 500 to almost 2 billion years, ten eras of hundreds of millions of years, 22 periods of 30 to 80 million years, 38 epochs from 13 to 35 million years and 98 ages covering the last 542 million years of the Phanerozoic eon. *Table 1.2* shows a simplified version[11].

The Moon was formed by accretion of debris derived from the collision of the proto-Earth with another protoplanet named *Theia (Halliday, 2000)* about 4.510 billion years ago *(Barboni, 2017)*; it was roughly the size of Mars and shared our same orbital space.

Dr. William K. Hartmann and Dr. Donald R. Davis were the first to suggest this leading modern hypothesis of the moon's origin in 1975 *(Hartmann, 1975)*.

---

[11]   The Table does not show the seven eras and ten periods of the Precambrian; 29 epochs that exist between the Cambrian to the Cretaceous are omitted, avoiding subdivisions such as "early, middle, and late," thus making it easier to understand. Finally, the 98 Phanerozoic ages that would have generated a highly variegated table were not included either.

Supereon	Eon	Millions of years
	Phanerozoic	541 ± 1
Precambrian	Proterozoic	2,500
	Archean	4,000
	Hadean	~4,600

Eon	Era	Period	Epoch	Millions years
Phanerozoic	Cenozoic	Quaternary	Holocene	0.011784
			Pleistocene	2.588
		Neogene or Tertiary sup.	Pliocene	5.332
			Miocene	23.03
		Paleogene or Tertiary inf.	Oligocene	33.9 ±0.1
			Eocene	55.8 ±0.2
			Paleocene	65.5 ±0.3
	Mesozoic or Secondary	Cretaceus		145.5 ±4.0
		Jurassic		199.6 ±0.6
		Triassic		251.0 ±0.4
	Paleozoic or Primary	Permian		299.0 ±0.8
		Carboniferous	Pennsylvanian	318.1 ±1.3
			Mississippian	359.2 ±2.5
		Devonian		416.0 ±2.8
		Silurian		443.7 ±1.5
		Ordovician		488.3 ±1.7
		Cambrian		541.0 ±1.0

Table 1.2   Simplified Earth Geological Timescale
Modified from The International Commission on Stratigraphy 2022.

The colossal frontal collision at ~4 Kms/sec caused the disintegration of Theia and enormous destruction of the Earth, causing the inclination of its axis[12] *(Connely, 2016; Young, 2016; Tanner, 2019)*. From that moment, the necessary conditions were given so the year's seasons could exist.

Another effect our satellite has since its formation due to its gravitational capture by the Earth is that it has a synchronous rotation, which means that its rotation speed coincides with the orbital translation speed, which is why we always see the same face of the moon.

Also, as a result of this collision with Theia[13], the Earth's original atmosphere made up of hydrogen and helium escaped into space. The Moon was integrated by accretion of the debris only 24,000 kilometers (15,000 mi) from the Earth's surface. When it completed its formation, it covered 250 times more area in the sky than it currently occupies and reflected hundreds of times more sunlight. Hence, it was impossible to observe any stars but can read with enough light at night. At that time, the Earth's angular momentum[14] and its massive satellite so close caused a very accelerated rotation. A full day had only 5 hours (so a year would be made up to ~ 1,800 days), and the Moon revolved around it every 48 hours. The earth's surface was largely magma, and meteorites continuously hit. In addition, the lava oceans were affected

---

[12] The inclination of the Earth's axis varies every 41,000 years from 22.1 ° to 24.5 °. Nowadays it is 23.437 ° = 23 ° 26' 12.3".

[13] Theia, Teia, or Thea in Greek mythology was a Titaness who married the Titan Hyperion, with whom she had three children: Helios (sun), Selene (moon), and Eos (Aurora). Thus, Theia is the mother of the Moon *(Halliday, 2000)*.

[14] Angular momentum is a physical quantity like impulse energy, generally rotational. The most common example occurs when an ice skater begins to turn with her arms extended to the sides of her body as she flexes her arms and brings her hands towards her chest; her rotation speed increases, although the angular momentum is preserved. The explanation is that the angular momentum is the product of the mass, the distance to the center squared, and the angular velocity—the latter increases as the distance decreases.

by the Moon's gravity, causing tides hundreds of meters high; the outlook must have been terrifying.

Besides Hartmann's shock theory of Theia for the Moon's formation, another hypothesis of its origin has been developed, such as that of multiple impacts *(Rufu, 2017)*. However, none of them have surpassed that of the single impact in computer simulations. Nevertheless, given the interplanetary conditions that existed 4.42 billion years ago, with abundant debris in Earth's orbit, a combination of both theories may have occurred *(Stevenson, 2014)*.

A kinetic characteristic of the Moon is its orbital distancing from the Earth, although at a decreasing speed. The first hundred million years at a rate of tens of meters per year, and currently, each year, it retreats at an average of 3.82 cm (1½ "), which has determined that its current average distance is 384,400 kilometers and that its orbital period has increased from 48 hours to 27.32 days.

Shortly after the impact, on our Earth, due to the environmental characteristics with surface temperatures exceeding 1,000 °C, the densest materials gradually migrated inwards due to gravity, and the lighter ones toward the surface. As a result, the gases released from the molten rocks formed a second atmosphere constituted not only by hydrogen and helium but also by water vapor, methane, and carbon monoxide.

Because of this migration process, the structure of the Earth was constituted of three concentric layers: crust, mantle, and nucleus *(Alberto, 2013)*, with a chemical composition consisting of iron 34.6%, oxygen 29.2%, silicon 15.2%, magnesium 13.9 %, sulfur 2.9%, nickel 1.8%, calcium 1.5% and the rest of the chemical elements of the periodic table make up the remaining 1.2%, sodium, and chlorine. However, even though iron is the most abundant element, almost everything is found in the nucleus (which is why we have a magnetosphere that protects us from the solar wind); with a diameter of 6,972 kilometers, it is a bit bigger than Mars. Therefore, oxygen is the most abundant terrestrial element in the crust, constituting ~47%, mainly combined with silicon, aluminum, and calcium *(Morgan, 1980)*.

The Earth has a density of 5.51 g / cm^3. It is the densest planet of the eight planets that make up the solar system, with a mass of 5.97 septillion kilograms (5.97 x 10^{24}) and an approximate hydrosphere of 1,388 sextillion Kg (1.38 x 10^{21}), equivalent to 0.023% of the total mass. The land surface has approximately 510 million km^2 (317 million sq mi), and 362 million km^2 (225 million sq mi) is covered with water, equivalent to 71%. Due to its composition, 97.5% is saline, and the remaining 2.5% is fresh (70% is frozen). The water in all the oceans contains ~35 grams of salt per liter. Its origin is because the dense primitive atmosphere (formed by the mantle's vaporization after the Moon's formation) had a temperature of 1,200 ° C with a high concentration of gasified water, sodium, and chlorine sulfides in smaller quantities. When the temperature dropped enough, these elements condensed with the water that fell for millions of years through the rain, forming the seas and oceans of the whole world *(Zahnle, 2010)*.

A few years ago, an astonishing discovery was published from a strange diamond found in a mine in the municipality of Juína in Brazil. It formed approximately 515 km (320 mi) deep within the earth's mantle. This diamond had a peculiarity; 1% of its weight was made up of water *(Pearson, 2014)*. This gem confirmed the suspicion since 1987 that there was water in the mantle *(Smith, 1987)*, but it had not been conclusively proven. However, other studies after Pearson have confirmed the existence of water in the so-called *Mantle Transition Zone*, located between 410 and 660 km (255-410 mi) deep.

On average, 1.5% of the weight of this enormous area could be made up of water *(Hongzhan, 2017; Tschauner, 2018)*. If these observations are fully confirmed, they would have an extraordinary impact on understanding our planet's structure, especially concerning the total volume of its hydrosphere. However, this fact is more significant; recently, evidence found water circulation between the seabed and the mantle. Now we know that it permeates even to the lower mantle. So, it is estimated that the volume of water contained in the entire mantle is 2.3 to 3 times greater than the water contained in all the world's oceans, rivers, and lakes *(Ohtani, 2020)* because the mantle represents 84% of the total volume of the Earth which is equivalent to one trillion Kms^3 (1.08321 x 10^{12} Kms^3) *(Nacagawa, 2019)*.

Nevertheless, even if the water were only present in the *"Transition Zone,"* its total mass would be 20% to 100% of the terrestrial oceanic mass *(Ohtani, 2018),* but in the molecular form called «Ice VII» In this form of water, the oxygen atoms are arranged in a cubic shape. In contrast, in the molecular structure of «Ice I$_h$, » which is ordinary ice, the arrangement of the oxygen atoms is hexagonal *(Fig. 1.8).* The structural difference is due to the enormous pressures (> 3 Giga-pascals equivalent to 30,000 atmospheres) that require the production of this Ice VII as it happens at the bottom of the ocean of Jupiter's moon Europa.

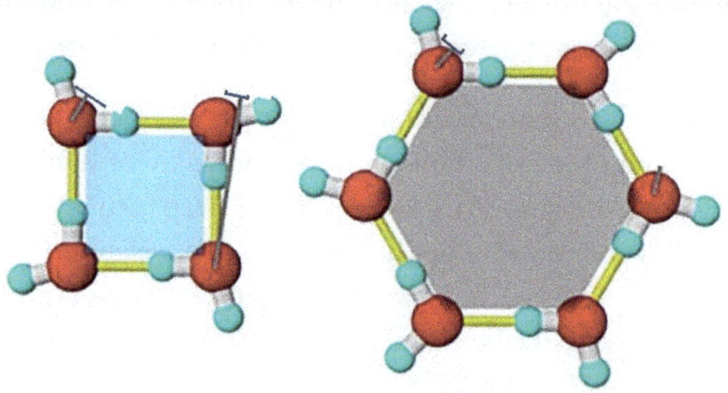

Figure 1.8   Ice molecules: Left: Ice I$_h$ Right: Ice VII.

Modified and adapted from Masakazu Matsumoto. American Institute of Physics

Now, precisely concerning this vital component, one of the most frequently asked questions is: where did the water come from on our planet? The answer is not so complicated if we remember that in addition to the synthesis of chemical elements in the cores of stars, other heavier chemical components of the periodic table are produced during supernova explosions. So, naturally, organic and inorganic molecules include water, so water is widely distributed in the Universe.

The protoplanetary disk from which the Earth was formed contained abundant water; as a result, all the planetary bodies had greater or lesser amounts of water. For example, Europa and Ganymede are two of the four Galilean moons of Jupiter; each of them probably contains more water than our planet.

In Jupiter, Saturn, and Uranus, the amount of water in their dense atmospheres is dozens of times higher. On our Earth, the primary water source most likely arose from the collision with Theia since the composition of this protoplanet was 50% in the form of water ice *(Dambeck, 2012; Gerrit, 2019)*. Thus, a few million years after the earth's formation, the water was constituted on its surface by atmospheric condensation due to the intense volcanism with the consequent formation of storms that lasted millions of years, forming the primitive oceans. An additional water source for Earth occurred due to heavy meteor and comet rain over 300 million years during the late heavy bombardment 4.1 - 3.8 billion years ago.

Closely related to this information, a surprising discovery was made recently: approximately 40% of the water on Earth is older than the Sun. This information is based on its high deuterium concentration, typical of the water formed in interstellar space and not inside the solar system *(Cleeves & Gibney, 2014)*.

Discoveries about the nature of our planet are occurring very fast. Yet, it is undeniable that we currently know more about the surfaces of the Moon and Mars than about the constitution of the bottom of our seas and oceans. Thus, the horizon of exploration on Earth is vast, and extraordinary discoveries await us in the future.

Well, so far, we have summarized the most important events that occurred from the beginning of time to the formation of the Earth. Now we will continue with an exciting topic —the origin of life.

Life

"What is life? A frenzy.
What is life? An illusion, a shadow, a fiction
… All life is a dream, and dreams… dreams are."

"Life Is a Dream"

—Pedro Calderón de la Barca.

Introduction

We will begin this chapter by saying that life is a consequence of the evolution of the Universe, and it is not by chance that, as part of it, we exist in this stage of the Cosmos. The Anthropic Principle, which we will talk about later, indicates that we are situated in the proper Universe at the right time. It seems that these arrogant assertions are overwhelming due to their level of evidence. That is, we live on a planet full of simple and complex life. Even in the face of so much diversity, we face problems such as knowing if some microscopic entities are alive or not, as is the case of prions which are infectious protein agents that produce diseases of the central nervous system.

However, life seems to be much less common than we imagine in the Universe, or at least we still have no news that it exists elsewhere *(McKay, 2014; Lingman, 2019)*. For example, in our solar system, none of the other seven planets with their 210 moons, including our *(Greshko, 2019)* nor the five dwarf planets, have shown concrete evidence of life. Moreover, it is not sure that any of the more than 5,200 exoplanets discovered so far have it *(Extrasolar Planets Encyclopaedia, 2022)*. However, it is feasible to exist in 12 of 29 selected planets in the stellar habitable zone[15]

---

[15] The circumstellar habitable zone is an extrapolation of the conditions of the Earth and the Sun. It refers to the theoretical area surrounding a star within which a rocky planet or satellite has sufficient atmospheric pressure (611 Pascals or higher) and range temperature so that there can be liquid water on its surface. Therefore, its conditions are favorable for the maintenance of life. This area has a variable distance according to the type of star. In stars like the Sun, this area is at a distance of around one astronomical unit (0.95-1.37 AU), which is equivalent to approximately 150 million kilometers (93.2 million miles), a star with 25% of the Sun's luminosity has a habitable central zone at about 0.5 AU. A star with twice the luminosity has a central zone around 1.4 AU.

The Kepler space observatory mission that began on March 6, 2009, and ended on November 15, 2018, was explicitly intended to find exoplanets with life possibilities in Cygnus's constellation. It discovered 4,000, and only twelve of them had some chance. In May 2013, due to equipment failures, the methodology had to be changed to continue the search for planets later. Through it, 381 new exoplanets were discovered; there are one or two with some potential for life.

Furthermore, in November 2013, NASA published that the number of Earth-like planets in the Milky Way is between 10 and 40 billion, based on the data obtained. With these data, the chances of finding life increase to levels never imagined 20 years ago.

Although we still do not have reliable data on extraterrestrial life, there is hope to find through new research projects like TESS (Transiting Exoplanet Survey Satellite), which until April 2022, has identified 5,488 candidate exoplanets, of which 205 have been confirmed *(NASA, 2022)*. Furthermore, the successor to Hubble, the James Webb Space Telescope, launched on December 25, 2021, and the Wide Field Infrared Survey Telescope "WFIRST" (or Nancy Grace Roman Space Telescope —to be launched in 2027, will be the first devices to investigate the dark matter directly and also will discover new planets that allow life on their surfaces.

On 11 July 2022, in a NASA livestream, US President Joe Biden released the first official image from the James Webb Space Telescope (JWST), showing a new image of a galaxy cluster SMACS0723 – a deep field into the distant Universe as it appeared 4.6 billion years ago. The view of the early Universe toward the southern constellation Volans was achieved in 12.5 hours of exposure with the Near InfraRed Camera.

Thousands of galaxies, including the faintest objects ever observed in the infrared, have appeared in Webb's view for the first time. This slice of the vast universe covers a patch of sky approximately the size of a grain of sand held at arm's length by someone on the ground. *Figs. 2.01 and 2.02*

Figure 2.01   JWST First Deep Field.
Credit: NASA, ESA, CSA, STScI. Public Domain.

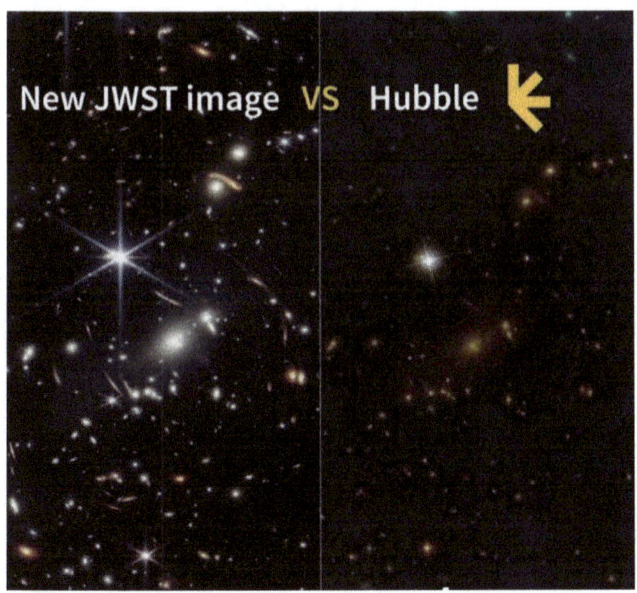

Figure 2.02   Same image taken by JWST and Hubble ST.
Modified from: Science alert. Public Domain.

On the other hand, the *SETI* (Search for Extraterrestrial Intelligence) project aims to explore and understand the origin and nature of life in the Universe. It began its search in late 1984 through radio telescopes calibrated to detect signals that can be interpreted as originating from intelligent life, just as we do when we emit electromagnetic spectrum waves into space from radio and television stations worldwide. Even to increase the possibilities of analyzing these signals, a global invitation was made from 1999 to March 2020 so that all those interested could collaborate with their home and office computer equipment through internet tools provided by SETI (Quest, Home, and Stars), through the Alien Telescope Array (ATA) so that all participants are aware of the received signals and can examine them.

However, until 2022 they have not received any emission from space to be interpreted as produced by intelligent beings. Nevertheless, the fact that we are not sure of extraterrestrial life is not indicative that it does not exist. As Carl Sagan mentioned, citing one of his favorite phrases, *"absence of evidence is not evidence of absence,"* after all, we have only explored the 0.00001% of all rocky planets in our galaxy.

Life is a complex and fascinating subject; its significance and meaning are fundamental in religion and philosophy, as are its origin and evolution for biology and cosmology. We will start this section with some basic concepts, progressing towards more complex and controversial aspects that will make this an exciting journey.

When touching on life, we face two initial problems: the first is to define it, and the second is to understand how it came to be. So, let us start with the elusive definition of life.

Concepts and Definitions of Life

There are more than 120 definitions of life in the literature *(Trifonov, 2011)*. For several years it has been insisted that the complexity of the meaning of life involves so many components and approaches that its definitions should be open and not limited to our current state of knowledge. Specific authors suggest that perhaps what is required is a frame of reference rather than a precise definition *(Tirard, 2011)*. We are trying to understand the Universe's most beautiful and essential component.

Some dictionaries and encyclopedias define life as those conditions that distinguish animated beings from inorganic matter; these include the capacity for growth, reproduction, functional activity, and continuous changes that cease with death.

In 1944 a book was published that was a milestone for the scientific world. This work entitled —*What is Life?* was issued based on the lectures given by the Nobel Prize in Physics Erwin Schrödinger a year earlier at Trinity College in Dublin. In his conferences, he defines life as a process of interaction between matter and energy based on the laws of Physics and Chemistry in a unique and complex way, giving rise to the laws of Biology.

Life does not create or destroy energy; it only transforms it, faithfully obeying the Law of Conservation of Energy or First Law of Thermodynamics. In the same way that every living being fulfills the Second Law, which indicates that, although the energy in the Universe is constant, it inevitably degrades, that is, it increases in "entropy," which is a term associated with disorder, randomness, or uncertainty *(Cox, 2013)*.

In the beginning, the singularity from which the entire Cosmos arose contained pure and homogeneous potential energy. Immediately after the Big Bang began the entropy process by degrading mainly into radiation, kinetic energy, and heat energy, which are types of energy of inferior quality to the original energy of the singularity. As part of the Universe, living organisms take chemical energy from food and dissipate it in heat.

They also advance individually in time in an entropic way, which causes them to age and eventually die. However, the life, Schrödinger wondered, how is it possible that life remains and progresses to more complex and orderly forms while the entire Universe advances from its origin continuously towards chaos and entropy?

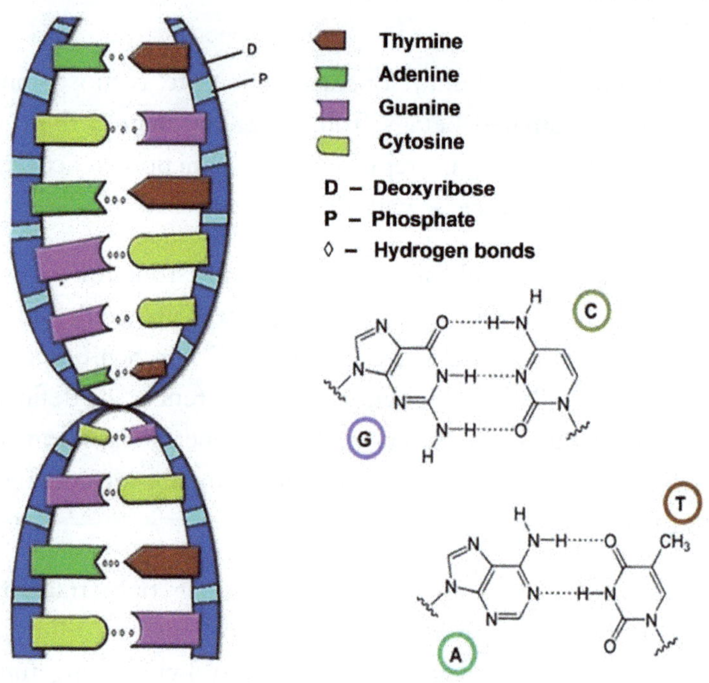

Figure 2.1 Structure of DNA / G≡C Mating with three hydrogen bonds. A = T pairing with two hydrogen bonds.

Modified from IGCSE Biology / Yikrazuul. Public domain

The answer to this paradox, he said, is found in the ability of the living beings to transmit the order to the next generation through genetic codes, which are the mechanisms by which life is renewed from generation to generation, allowing it to way to evolve in complexity. This work profoundly influenced Wilkins, Watson, and Crick in their academic lives, prompting them to discover the DNA molecule for which they won the Nobel Prize in Medicine[16] in 1962.

The DNA molecule (deoxyribonucleic acid) is the basis of the genetic code Schrödinger referred to in his conferences.

The structure of DNA is composed of a double helix skeleton made up of a five-carbon sugar (deoxyribose) linked to a phosphate group on the one hand and linked to pairs of nitrogenous bases or nucleobases: adenine (A), cytosine (C), guanine (G), and thymine (T), in turn, linked together by hydrogen bonds: three bridges for CG and two bridges for AT. Since the junctions of these four bases are specific, C only binds to G, and A only binds to T *(Fig. 2.1)*. During reproduction, the double helix separates, leaving two individual helices, and the corresponding complementary letters are synthesized immediately, finally constituting the two new double helices.

These are the basis of DNA replication, cell division, and what allows our cells to grow and multiply. RNA (ribonucleic acid)[17] is made up of a pentose (ribose) that has one more oxygen molecule than deoxyribose and contains a nitrogen base different from DNA —uracil— instead of thymine, which also binds to adenine *(Figure 2.2)*. The messenger RNA carries out the transcription of the DNA

---

[16] The English chemist Rosalind Franklin contributed decisively to discovering the structure of DNA by obtaining images of it by X-ray diffraction. Wilkins stated that Franklin should have also been awarded the Nobel Prize in Medicine in 1962 (symbolically already who had died in 1958). However, neither Watson nor Crick mentioned her when receiving her awards.

[17] As we will see later, everything indicates that the genetic code was RNA and not DNA at the beginning of life.

information so that later the transfer and ribosomal RNAs translate protein synthesis.

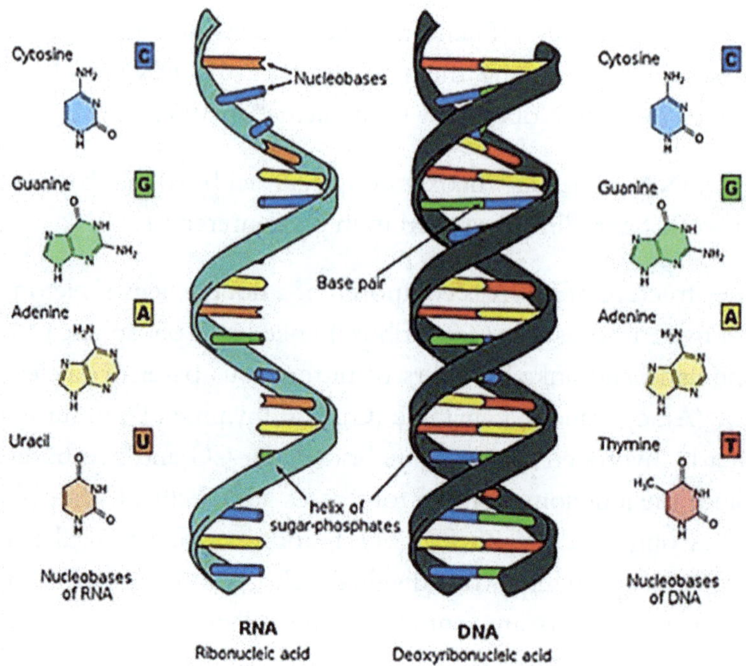

Figure 2.2 **Representation of RNA and DNA with their respective nitrogenous bases.**

Credit: Darryl Leja. National Human Genome Research Institute. CC BY 3.0.

In addition to the four natural bases, artificial DNA nitrogenous bases have been synthesized by the so-called –polymerase chain– reactions, which are extended genetic alphabets *(Yang, 2010)*.

They receive the generic name *"Artificially expanded genetic information systems"* (AEGIS). They are non-natural forms of DNA that increase the number of nucleotides building blocks that replicate independently of natural systems. These nucleotide systems do not bind to natural DNA. There are four artificial base pairs in addition to the two natural C-G and A-T, with the Z-P pair being the most used *(Fig. 2.3)*.

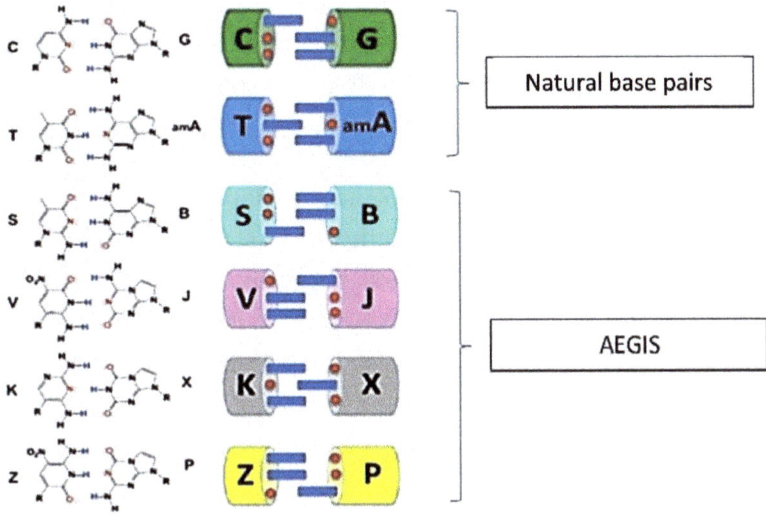

Figure 2.3   Expanded Genetic Information Systems (AEGIS).
Modified from: Sefah, K. et al., (2014) *PNAS* 111 (4), 1449-1454

Advances in genomics, genetic engineering, and synthetic molecular biology techniques are astonishing[18].

However, let us go back to the definitions of life —For many years, it has been known that metabolism is the central feature of life. This conclusion has led to the consideration that viruses are non-living entities.

At the beginning of the eighties, Varela and Maturana defined life as an « *autopoietic system,* » which means that it is an entity constituted by an internal process of self-maintenance and self-generation translated as metabolism and reproduction *(Varela, 1974 & 1981).*

---

[18] The Polymerase Chain Reaction is a molecular biology technique that relies on the natural property of DNA polymerase to replicate DNA strands and thus obtain many copies of a particular DNA fragment. This technique amplifies any DNA fragment, making it much easier to identify bacteria or viruses, identify people or corpses, and conduct scientific research on DNA.

On the other hand, life is a thermodynamically open system from the physical point of view. It means that it exchanges matter and energy with its environment.

In 1992, the board of the NASA Astrobiology Program asked its scientific group to develop an operational definition of life that could be applied to any possible form in the Universe. After analyzing several options, the winning was the one suggested by Gerald Joyce, who states: « *Life is a self-sustaining chemical system capable of carrying out Darwinian evolution.* » Without doubt, a simple, beautiful, and intelligent description.

By unifying the three exposed definitions, we conclude that: *"Life is a complex autopoietic chemical system, thermodynamically open, capable of experiencing Darwinian evolution" (Tirard, 2011),* and we would have to add that not only Darwinian but supradarwinian; and what does this mean? Well, it means that humanity today has taken artificial selection to highly advanced levels. We can modify genetic sequences and clone plants, animals, and even human beings. Moreover, we are beginning to create artificial life. So, some astrobiologists think that it is likely that extraterrestrial beings have already done all of this on their planets and have even achieved far superior results. In such a way, the definition of life in the Universe must be extended to the level of supradarwinian evolution if we want to improve the ability to identify it in other places in our galaxy *(Benner, 2010).*

One of the ways to understand life is by knowing what its essential characteristics are. Complexity is probably the best representative of these characteristics; everything that exhibits life must be complex. For example, it grows, reproduces, has metabolism, has an information storage system, and evolves by natural or artificial selection. However, there are complex non-biological systems, like crystals, that grow and replicate, and some computer programs are also stored, and they even infect software systems through computer viruses. Another example is that 3D printers can reproduce themselves by manufacturing the parts that compose them one by one; on the other hand, fires have a certain degree of metabolism since oxygen is consumed during the ignition process of objects, and energy is released *(Cockell, 2013)*. Those who have tried to define life from a single perspective may be right, but only partially.

Historically Empedocles (495-435 BC), creator of the Greek school of materialism, asserted that the world is composed of four elements: —earth, water, fire, and air; and that life has a soul made of fire. Leucippus and his student Democritus of Abdera (470-380 BC), initiators of the *mechanistic atomistic theory*, believed that all matter, including life, was made up of tiny indivisible particles called atoms. Life had a particular type of atoms scattered throughout the bodies of the living beings, providing them with vital characteristics.

Another ancient concept concerning life is that of "spontaneous generation," whose origin precedes Aristotle (384-322 BC), who called it « *abiogenesis* » and remained valid for more than 2,200 years until the nineteenth century. In the 18th century, the Enlightenment developed the Vitalism theory, which indicated that life contains a vital force that transforms death or inorganic material into living matter. One of the recipes known by tradition for a century before mentioned: « *in a container, add underwear and grains of wheat, wait 21 days for the "vital force," and you will get mice.* »

In 1707, the German physician George Ernst Stahl published his book *" The True Theory of Medicine,"* emphatically asserted that all living organisms are "special" because they are not governed by physical laws but by-laws of a different character. In direct opposition to this idea, the Dutch Hermann Boerhaave studied the functioning of the human body and, together with Friedrich Hoffmann, showed that all its activity obeys the laws of physics and chemistry. The first position is called *"vitalist or animist"* and the second *"mechanistic."* The French philosopher and mathematician René Descartes was a prominent defender of the mechanistic stance.

Francesco Redi (1626-1697) Italian physician and physiologist, was the first scientist to show that spontaneous generation was false. In

1668 he experimented with placing raw beef in three jars, one jar left uncovered, another covered with gauze, and the last with a hermetic cork, and left them in the open air. After several days, the fly larvae only appeared on the meat in the uncovered jar but not on the corked or gauze-capped jar (although larvae did appear on the gauze). In all pots, the meat was rotten. He also performed the same experiment with fish obtaining the same results. Despite the evidence, he preferred not to promote his investigation for fear of reprisals from the Inquisition.

However, the issue of the proliferation of microorganisms observed even though the jars were hermetically covered was still pending. So, Luis Pasteur, in 1861, based on an experiment that Lazzaro Spallanzani had carried out 92 years earlier, used jars with gooseneck sealed with fire, prevented their contents from coming into contact with the environment, boiled a culture medium, and sealed them.

In the experiment, Pasteur broke the neck of one of them, which caused the microorganisms suspended in the air to contact the culture medium. A few days later, it became cloudy because of these microbes, unlike the flask sealed in which the culture broth was kept intact, thus demonstrating that the spontaneous generation was false. As Pasteur repeated the experiment several times, some of these flasks are still preserved in the Luis Pasteur Museums in Paris and the Science Museum in London —all remain sterile nowadays.

However, the falsehood of the last defense of the Vitalism theory had yet to be proved. The defenders of this theory claimed that the heat in the air destroyed the Life Force. So, the *coup de grâce* to this theory was given by the British physicist John Tyndall in 1876 when he built special chambers, the inner surface of which he coated with glycerin *(Fig. 2.4)*. The closed chambers were left untouched for several days until a beam of light passing through the side windows indicated that all the floating matter in the air had settled and adhered to the glycerin surfaces.

According to Tyndall, the air was then « optically pure. » Under these conditions, all kinds of previously sterilized organic liquids such as urine, broth, and vegetable infusions could be exposed to the air in the chamber, which remained unchanged for months; in other words, the optically empty air was also sterile. The conclusion was that the power of the atmosphere to engender bacterial life is related to its dust content and that many of the microscopic particles that float in the air are formed by microorganisms or are transporters of them. Furthermore, the air inside the chamber was not heated, so the life force was not being destroyed. Putrefaction did not occur in dust-free air *(Yancovic, 2010)*. Thus, thanks to Pasteur and Tyndall, the idea of spontaneous generation was eliminated, and the Biogenesis Theory was verified.

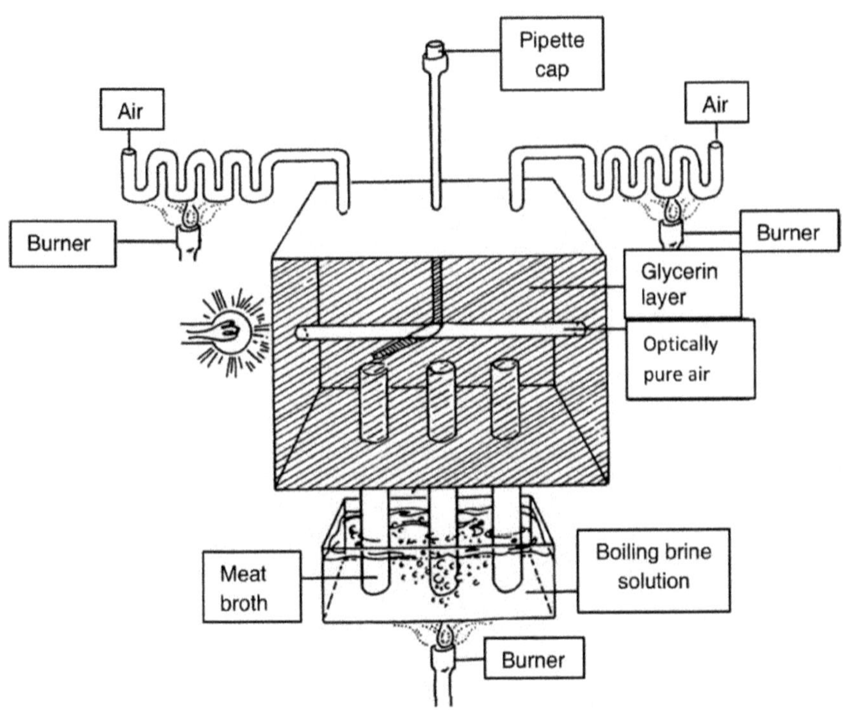

Figure 2.4 Tyndall's Chamber.
Public domain.

In 1939, the German pathologist Rudolph Virchow developed the *"Cell Theory,"* whose main statements indicate that every cell proceeds by division of another already existing and that cells are the anatomical and physiological units of living beings whose genetic material is the necessary element so that the mother cell can pass vital information to the offspring cell.

In conclusion, we learned through these examples that eliminating misconceptions requires inventiveness, time, and much effort for men and women committed to getting to the truth. Generalized ideas, although erroneous, are complicated to eradicate. Ignorance is very daring and is supported by ancient traditions, assumptions, and beliefs whose origin is magical thinking.

So far, we have broken down some concepts and definitions of life; now, the second problem is to answer the question: How did life begin?

Origin of life

The principle of life is intrinsic to cosmogony in such a way that, to explain it, we must return to either of two paths: the creationist or the expansionist-evolutionist. As we have reviewed so far, the precise understanding of how it happened in creationist philosophies is not essential. Still, it involves believing in a well-crafted story and accepting it as true by incorporating acts of faith. On the other hand, in the expansionist-evolutionist theory, life arises from the time, matter, and energy of the Universe and the laws that govern it from its origin. One of the most significant differences between creationist and evolutionary theories of life results in biological complexity. While in the creationist, plants, and animals, including human beings, appear instantaneously as such by a mere act of the will of a god or gods, in the evolutionary it results from a slow evolutive and adaptive process stemming from the *"Phylogenetic Tree of Life"* that we will review later. The appearance of life or biological self-organization is an inevitable emergent process when the right conditions are present for the complex dynamic systems that make it up to development *(Friston, 2013)*.

Thus, the fundamental concept of life includes 1) a process of interaction between matter and energy based on physics, chemistry, and biology laws, and 2) an evolution by natural or artificial selection whose characteristics include growth, metabolism, and reproduction. Explained in this way, the essence of life is understandable to us; however, how it began is a different matter.

The Russian biochemist Alexander Oparin in 1923 published his book *"The Origin of Life,"* in which he postulates that in the primitive

Earth, due to the solar ultraviolet radiation and the electrical discharges of the constant storms, the inorganic molecules of the atmospheric gases gave rise to prebiotic organic molecules. He called this set of reactions *prebiotic synthesis*, and these molecules were amino acids and nucleic acids. According to Oparin, these molecules became trapped in shallow pools formed on the early ocean's shores, which he called *primordial broth*.

As they concentrated, they diversified, forming complex biopolymers (e.g., nucleosides, peptides, polynucleotides) that then evolved into functional multimolecular systems that were finally "alive" *(Oparin, 1923)*. Both Oparin and the British John Haldane, with some differences between their approaches, postulated the origin of life as sequencing that was part of the Earth's evolution process *(Tirard, 2010)*. The basis of his reasoning is called the *"Prebiotic Soup Theory,"* which is the synthesis of organic compounds from a reducing atmosphere.

At the beginning of the 20th century, the French biologist Stéphan Leduc, in his work *"La Biologie Synthétique,"* described that life arises as intermediate forms between the inorganic and the organic world. For example, rocks in the ocean have carbon compounds that combine with other chemical elements, mainly hydrogen, oxygen, nitrogen, phosphorus, and sulfur, giving rise to organic chemistry and biochemistry. However, having the building blocks of life does not mean that it originates from that simple fact. This claim evidenced Stanley Miller and Harold Urey's experiment in 1953 at the University of Chicago. Both researchers designed a device that simulated conditions in the early Earth's atmosphere *(Fig. 2.5)*.

The experiment involved subjecting a mixture of methane, ammonia, hydrogen, and water to electric discharges of 60,000 volts. This experiment formed a series of organic molecules, among which acetic acid, ADP-Glucose, and the amino acids —glycine, alanine, glutamic acid, and aspartic acid stand out. This experiment was crucial for reinforcing Oparin and Haldane's theory of the primordial soup for the origin of life. However, it is known that Earth's early atmosphere ~

3.9 billion years ago was made up of water, carbon dioxide, and nitrogen, with small amounts of carbon monoxide and hydrogen and almost no methane and ammonia *(Zahnle, 2010)*.

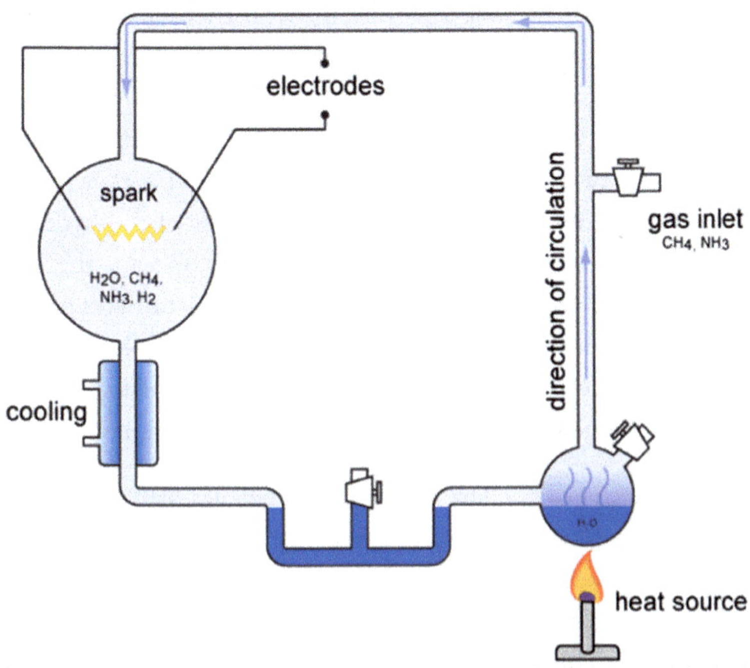

Figure 2.5 Miller and Urey experiment.

Credit: Carny at Hebrew Wikipedia. Aug 2006. CC BY 2.5

The Miller-Urey experiments were only a first step since they obtained simple biomolecules but did not form the necessary polymers present in all living organisms. So, Sidney Walter Fox and his workgroup at the University of Florida conducted a series of experiments between 1957 and 1965 mimicking the conditions in the vicinity of volcanoes and surface hydrothermal vents such as geysers and hot springs ponds. Through a process called thermal copolymerization of amino acids, they obtained different polypeptides such as proteinoids that agglomerate to form microspheres that exhibit certain cell-like behaviors.

For this reason, they proposed that this type of structure could be the ancestors of cellular organization, calling them *protocells.*

Investigations of the origin of life in hydrothermal vents continued by other researchers in the ocean's depths, where structures called chimneys or smokers were discovered in the Pacific Ocean by a team from the Scripps Oceanographic Institute in 1977, 2,450 meters deep. They identified two types of structures: *Black Smokers* that mainly cast iron and sulfides into ocean water, are the hottest of the hydrothermal vents, reaching temperatures from 60 ° to 460 ° C with the ability to grow to more than 10 meters in height, and the less hot and more petite *White Smokers* with emissions that look like white smoke because they do not contain the metals found in the emanations of black fumaroles but are rich in minerals mainly barium, calcium, silicon, and zinc.

In the depths of the sea where the fumaroles are, sunlight is non-existent, yet the researchers found many life forms that constitute the food chain's base. For example, *Archaea* and extremophilic *Bacteria* convert heat, methane, and sulfur compounds into energy through chemosynthesis. They even found a species of phototrophic bacteria at a depth of 2,500 meters, part of the *Chlorobiaceae* family, which uses the faint glow of magma at the base of black fumaroles for photosynthesis. These are the first organisms discovered in nature with a natural light source other than the sun to conduct photosynthesis *(Beatty, 2005).*

As we see throughout history, various scenarios have been located for the emergence of life: Charles Darwin placed it in a warm pond where organic molecules existed; Oparin, Haldane, and Miller in the primitive atmosphere during electrical storms, where combinations of organic molecules produced in the air were deposited due to the rain in proper places for their later evolution towards more complex molecules counting on the existence of other molecules pre-formed and with the help of solar radiation, especially ultraviolet rays.

Furthermore, Fox locates the origin of life in superficial hydrothermal vents near volcanoes, giving the possibility of bio-polymerization in those places.

Also, as we have just seen, the oceanic vents are the other scenario where life could arise due to the temperature gradients and hydrogen ions and the differences in concentration of various salts in seawater. All these make an ideal place for complex organic and biochemical chemical reactions.

Finally, another framework from which life could originate was some saltwater beds, where stromatolites were formed *(Fig. 2.6)*. Structures are molded by accretion and biochemical stratification of bacteria, specifically cyanobacteria. There are 3.7-billion-year-old stromatolites in Greenland.

Figure 2.6 Stromatolites in Shark Bay, Western Australia.
Credit: Peter Rusell, University of Waterloo.

For more than 50 years, the Miller-Urey hypothesis dominated the scientific environment and was replaced by those of the hydrothermal vents at the bottom of the sea. Finally, that seemed to be the winner in this complex debate on the generation of life.

Recently there has been a different scientific current in which the origin of life is not located in the depths of the ocean but based on the theories of Sidney W. Fox of superficial hydrothermal vents, but with some variants *(Deamer, 2015)*. The assembly of cell membranes occurs more readily in low-intensity ionic solutions with a minimal content of salts and divalent cations. That suggests cell life could have started in freshwater ponds of hydrothermal vents adjacent to volcanic islands (rather than submarine hydrothermal vents), where lipid-encapsulated polymers can be synthesized through hydration and dehydration cycles to form protocells *(Damer, 2020)*.

Life as a biophysical phenomenon associated with biochemical reactions requires energy to carry out these chemical reactions. Initially, the proton gradient gave the energy so carbon could combine with other elements to form organic molecules. Some of these molecules, such as ammonia, methane, water, carbon monoxide, and carbon dioxide, had already been created from the explosions of the first supernovae long before the formation of the Earth *(Hazen, 2013)*. On the other hand, cellular life arose by abiogenesis with elements of inanimate matter. Starting from the mineral kingdom, but different from the concepts of spontaneous generation handled in ancient times.

At the molecular level, clay forms large overlapping layers facilitating the construction of organic molecular complexes. Without clay, forming these molecules, such as amino acids and RNA, would have been difficult or impossible *(Fig. 2.7)*. Negatively charged mineral surfaces could also directly stimulate the formation of membranes. However, they only coat and catalyze in other cases, forming vesicular membranes encapsulating these mineral particles with catalytic properties *(Hanczyc, 2007)*.

In the same way as the clay layers, the layers of mica crystals in a network can perform similar functions. In both cases, the integration of polymersomes was composed of amphiphilic molecules[19] that would later form membranes, isolation, and compartmentalization *(Jheeta, 2017)*.

In addition to this source of organic compounds, those reactions required liquid water and energy to drive these chemical reactions. This way, a molecule emerged and could make copies similar to itself, "the first replicator" with a constitution probably similar to RNA.

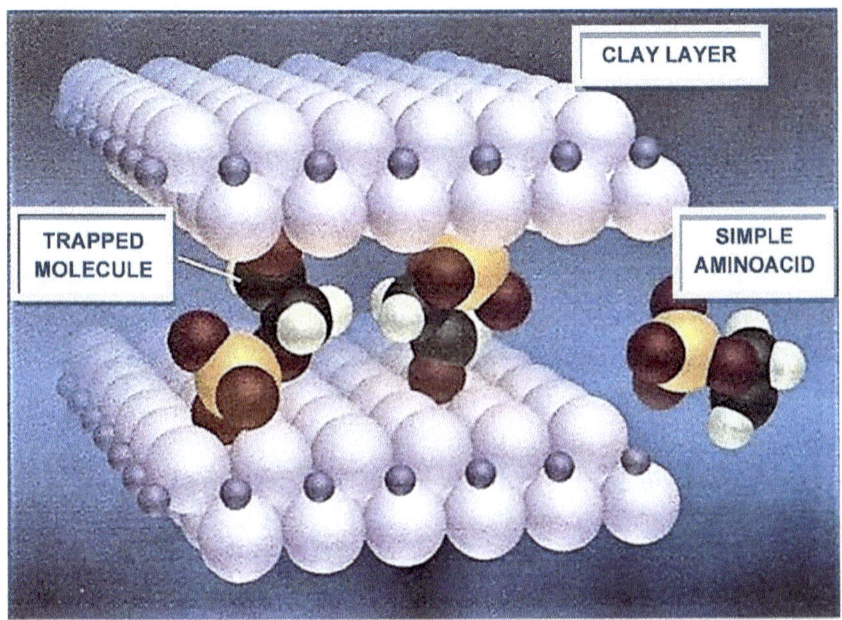

Figure 2.7   Formation of organic molecules between layers of clay.

Modified from: Origin of Life on Earth. College of Santo Domingo.

Another concept developed a few decades ago to explain the emergence of life is the *"Dual Origin of Life Hypothesis"* by Freeman Dyson. He postulated that from an initial world of proteins, a world of

---

[19] Amphiphilic: A chemical substance with both hydrophilic (affinity for water) and lipophilic (affinity for fat) properties.

RNA eventually originated as a by-product of increasingly sophisticated metabolism[20].

Thus, the world of RNA, which begins as an obligate parasite of the world of proteins, ends up constituting a system of cooperation and replication. Hence, the constitution of life as we know it today *(Dyson, 1985)*. This original concept, with some variations, is currently accepted *(Koonin, 2006)*.

Recently, from the knowledge about membrane systems and RNA as initial life models, some essential characteristics of the primitive environment have been deduced. For example, RNA replication requires chemical energy and the division of vesicular membranes for mechanical energy. These requirements have led to the location of another probable origin of biology on the surface of a lake.

The prebiotic chemistry of concentrated cyanide gave rise to nucleotides, amino acids, and perhaps fatty acids. Cyanide molecules are the main constituents of this prebiotic chemistry in laboratory models. Cyanide was relatively easy to produce in the early oxygen-free atmosphere with ultraviolet rays and meteorite impacts. Another necessary condition arises from the RNA replication cycle, which requires moderate changing temperatures.

Only lakes in geothermally active zones provide that temperature-swinging environment - lakes similar to those in Yellowstone that can be cool, even ice-covered during winter, but contain numerous hydrothermal vents that emit jets of hot water. In such an environment, the protocells were occasionally swept away by these high-temperature currents, which could cause the temporary separation of the RNA strands.

---

[20] An Anglo-American theoretical physic mathematician. Wrote "The Origins of Life" in 1985 based on the Tarner Lectures he gave in Cambridge that same year. Dyson builds a mathematical model abstractly representing the transition from chaos to organized metabolic activity in a population of molecules.

Nevertheless, these protocells were quickly mixed in the cold-water zone before the heat destroyed their delicate RNA molecules. Thus, combining these favorable chemical and physical environments would be the most likely environmental scenario on early Earth for the origin of life *(Szostak, 2016).*

In November 2016, the "Network of Researchers on Horizontal Gene Transfer and the Last Universal Common Ancestor" (NoR HGT and LUCA) was established. This group determined by consensus the *"Hypothesis of Multiple Root Genesis."* This eclectic consensus offers a scientific environment of more freedom and respect to expose various theories about the origins of life without the favorite or the most respected, as has been happening in the last 90 years. In conclusion, nowadays, *«no one knows how and when the transition from inanimate world to life occurred. »*

The organic molecules form in the Universe in various places. For example, they are found in interstellar dust by the action of cosmic rays and the explosions of stars. On our planet, biomolecules are formed in the atmosphere, the sea, freshwater, near volcanoes, and meteorites' impacts.

Life on our planet currently occupies all possible niches, especially microbial life, the most abundant on Earth, so its extinction is almost impossible. This situation was described by the director of the Department of Astrobiology at the University of Edinburgh, Dr. Charles S. Cockell, in his book *"Impossible Extinction: Natural Catastrophes and the Supremacy of the Microbial World" (Cockell, 2003).* In the event of life disappearing, the extinction process would occur in the reverse order of its emergence. The last levels to succumb would be the most primitive and resistant, specifically the extremophilic bacteria. As their name implies, they can tolerate extreme temperatures, radiation, droughts, acidity, and alkalinity.

Energy for Life and Panspermia Hypothesis

The term proton refers to chemistry's hydrogen ion (H +). Since the atomic number of hydrogen is 1, when a hydrogen ion has no electrons, it corresponds to a bare nucleus that consists of a proton (positively charged and therefore a cation). So, the terms "protons" and "hydrogen ions" are synonymously in chemistry. The transfer of these hydrogen ions in an acid-base reaction is commonly known as a "proton cascade or gradient." An acid produces hydrogen ions in a solution (water), a proton donor, and an acceptor of a pair of electrons. A base is an acceptor of a proton and an electron-pair donor to produce hydroxide ions in water.

Protons (H+) are exceptionally chemical reactive and react immediately with the electron cloud of any available molecule. This reaction causes the electrochemical proton gradient from the more important place to the minor concentration area, leading to chemical energy generation. Proton transfer (H+) was the energetic key for the initial generation of life. The kinetics of the first chemical element formed after the Big Bang, hydrogen, is the origin of the energy of life, and carbon, one of the elements to be constituted shortly after, is the basis of living matter. The energy provided by this cascade of hydrogen ions moves the carbons to recombine with other elements and give rise to organic chemistry *(Cox, 2013)*. Thus, the nature of our Universe is inherent in life.

There could probably be another chemical element on other planets besides carbon capable of supporting the complex chemistry necessary for life. The most probable, postulated by some astrobiologists, is silicon (one of the most abundant elements on our planet) due to its quality of allowing up to four chemical bonds like carbon. However, it does not have the same ability to keep the bonds with other molecules such as hydrogen, nitrogen, or carbon in equilibrium, having a significant "preference" for oxygen, forming the silica or silicon oxide that constitutes the sand of all beaches and deserts of the world. It is also capable of creating other molecules, but they are highly unstable, like silane (SiH4), unlike its much more stable carbon-based counterpart, methane (CH4) *(Plaxco, 2011)*.

Another essential point to know is that in the Milky Way, for each silicon atom, there are seven carbon atoms *(Croswell, 1996)*, so it is much more likely that life, at least in our Galaxy, is based on carbon.

It is necessary to mention that this cascade of protons for the coupling of carbon with other chemical elements would not be carried out without the appropriate and vital environment for this to happen —it is the liquid medium constituted by the solvent molecule par excellence, "water." Water is a combination of the most common element in the Universe (hydrogen) and the most common element in the Earth's crust (oxygen) and is the main constituent of the structure of living beings. Experimental attempts recreate vital processes with other potentially biotic solvents such as ammonia, methane, and hydrogen fluoride. However, none of them have the necessary characteristics to allow the coupling of the molecules required to develop life, at least not on our planet.

Of course, the proton gradient was not the only source of energy to drive the origin of life. On the primordial Earth, there was intense volcanism, electrical storms, and cosmic radiation from the sun; thus, the first recorded photosynthetic activity dates back to one billion years after the formation of the Earth. Moreover, photosynthesis is a chemical "battery" of carbohydrates and atmospheric oxygen *(Eriksson, 1991)*.

Now, when affirming that life is part of the essence of the Universe, this leads us to two hypotheses: First, that must be very abundantly given that there are hundreds of trillions of rocky planets in the trillions of galaxies that make up the observable Cosmos *(Conselice, 2016)*. Second, although the origin of life on Earth could have occurred inside it, it is also likely that it came from other planets or satellites of our solar system or other systems. This last assumption or hypothesis is called Panspermia, whose name dates from Anaxagoras (500–428 BC), whose teachings spoke of the "seeds of life." In 1903, Swedish chemist Svante Arrhenius used panspermia to explain that some simple forms of life got trapped in debris ejected into space after collisions between planets and petite bodies and traveled dormant through space embedded in meteorites to other worlds, including Earth.

In the beginning, this hypothesis was well received and had the support of famous scientists such as Fred Hoyle, Francis Crick, and Leslie Orgel, among others. Nowadays, there is little controversy in the scientific community, and the general tendency is to give it little credit, given the overwhelming evidence to the contrary derived from recent observations *(Willerslev; Ponce de León; Bhargava, 2003 & Giulio, 2010)*.

However, certain extremophilic microbes can survive the initial hit of meteorite impacts. Therefore, we can assume that they could survive thousands of years traveling through space before reaching another planet. Mars, in particular, seems to be the planet from which life could reach Earth since it completed its formation several million years before. That is why its size is so tiny, while Earth and Venus continued to grow due to the accretion of more debris, which lengthened their prebiotic era *(Lui, 2011)*. On the other hand, Transpermia, a variant of Panspermia, indicates that life on Earth originated strictly within the solar system, while panspermia implies origin even from extrasolar planets.

In the Solar System's early days, the sun shone with only 70% of its current intensity. Despite this, the climates on Mars and Earth were more similar than they are now. Mars had a composite atmosphere of oxygen, methane, and carbon dioxide. These last two gases produced a greenhouse effect, so there were oceans of liquid water on its surface that remained for 800 million years (from 4.5 billion years ago to 3.7 billion years ago). In such a way, the life that had settled there could have reached Earth *(Cockell, 2015)*. However, the period of "Late Heavy Bombardment" of asteroids that affected Mercury, Venus, Earth, and Mars, which occurred 4.1 billion years ago and lasted for 300 million, probably destroyed the life that may have existed on both planets.

Since the formation of the Earth, several tons of rock have traveled from Mars, torn up by asteroid impacts on its ground, and traveled through interplanetary space before striking the Earth's surface. All of this happens because the escape velocity on Mars is only 5 kms/sec (3.1 mi/sec), representing less than half the escape velocity of our planet. However, only 340 identified meteorites from the red planet are currently known *(Meteoritical Bulletin Database, 2022)*, and none have obtained any microbial culture.

Comets are the other source from which life on our planet could have come since their main chemical composition consists of water and organic molecules such as carbon monoxide, methane, ethane, ammonia, acetylene, and methanol. The current scientific consensus is still open to all possibilities, although the extraterrestrial origin is remotely possible. It is well established that our planet had all the necessary elements for the emergence of life and the only advantage Mars had over us was precisely the time.

Regardless of whether or not life arose in our world, the how and not where is the point to be solved. Since we do not know much about the initial environmental conditions of Mars and almost entirely of any other planet or moon from which the seed of life could have come to ours, we must detail what we know based on the fact that life has arisen on our planet. The theory of panspermia could be the other way around, and our planet Earth could be the sowing of life on other planets and moons, as will happen in the distant future when we colonize other planets.

The Tree of Life

It is essential to include two indispensable hypothetical pieces of the prebiotic era to understand the origin of Biology: The *Protobiont* and the second is its derivative, the *Progenote*. A protobiont or protocell is a hypothetical precursor to single-celled life. These protocells were diverse compartmentalized systems of molecules, differing in composition from one another, and each represented a kind of natural experiment.

They consisted of a lipid membrane with abiotically formed proteins and nucleic acids inside and a simple energy generation and information transmission system. Due to the nature of the Earth more than four billion years ago, this protocell could have been anaerobic, hyperthermophilic, and chemolithoautotrophic. That is, oxygen was absent in its respiration, tolerated extreme hot and cold temperatures, and fed itself with the rocks' chemical elements *(Chen, 2006)*. However, as we will see shortly, the reality was slightly different, according to recent discoveries.

The second element, the progenote or first replicator, was a primitive entity that must have possessed a rudimentary and imprecise link between its genotype and phenotype *(Di Giulio, 2011)*. The relationship between genetic information and morphological and functional characteristics was still evolving. The limitations of its rudimentary translation mechanism give us the certainty that the progenote was a unique entity, unlike today's existing life forms. It lacked the current level of translation precision, so it could not synthesize normal-size proteins without introducing numerous errors. That means the progenitor could not have "modern" complex proteins; its proteins must have been small and had shifting sequences.

Consequently, the progenote enzymes may not have been as precise and specific as their current equivalents, which should have limited their control and definition mechanisms *(Woese, 1987)*. In the same way, their inaccurate replication mechanisms affected their survival. Any change due to important mutations, the next generation no longer appeared. At that point, the natural experiment was over. However, on other opportunities, some favorable changes achieved an improved replica so that this variety would become numerous and successful in its prevalence.

The currently most accepted sequence of how life originated describes that the protobiont evolved into a progenote. When it achieved a precise link between its genotype and phenotype, it formed a –genote– corresponding to LUCA *(Last Universal Common Ancestor)*. However, some authors consider LUCA was still a progenote *(Di Giulio, 2011)*. Regardless of this controversy, the last universal common ancestor gave rise to the —Phylogenetic Tree of Life *(Fig. 2.8)*, a cellular system of three domains: Archaea, Bacteria, and Eucarya *(Gogarten, 2016)*.

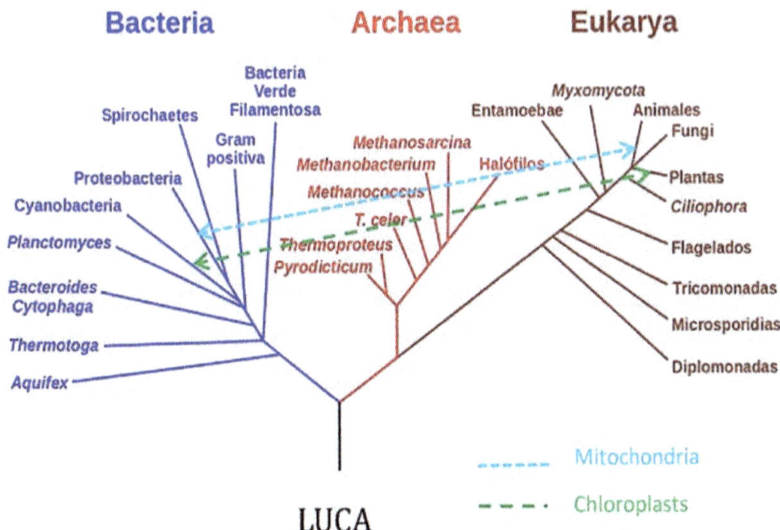

Figure 2.8 Phylogenetic Tree of Life by Eric Gaba (2003)
Modified from: NASA Astrobiology Institute. Public Domain.

In addition to these three branches from LUCA, adding viruses as a non-cellular molecular system to the origin of life is essential. As it is known, they were capable of infecting the three original cellular domains, being called due to this: archaeovirus, bacteriavirus, and eukaryavirus. Moreover, it has been suggested that the first viruses could exist before LUCA [21] or were contemporaries. It is very recent knowledge that the first DNA chain was viral. In such a way, viruses (retroviruses) induced the transition of cells with an RNA genome to transform into cells with a DNA genome, triggering the emergence of the three cell domains. Not only that, but many viral genes are present nowadays in all cells, including the nuclear, mitochondrial, and chloroplastic genomes. Patrick Forterre, head of the Department of Microbiology at the Pasteur Institute, cleverly points out that *"viruses infect the Tree of Life from the roots to the leaves." (Forterre, 2010; 2015).*

The LUCA was not probably a primitive single, hyperthermophilic prokaryotic cell but a complex community. A mixture of protoeukaryotic cells with an RNA genome, genetically redundant, adapted to a wide range of moderate temperatures, and morphologically and metabolically diversified [22] *(Di Giulio, 2011).* The latest research maintains that LUCA appeared 4.4 to 4.2 billion years ago *(Camprubí, 2019)* and gradually evolved, acquiring the level of complexity of a prokaryote. It gave rise to two superkingdoms ~ 4.2 billion years ago: the *archaea* and the *eubacterium.* Finally, *eukaryotes* emerged from the union of these two domains 2.7 billion years ago, resulting in the appearance of the cell nucleus.

---

[21]   S. Blair Hedges and Sudhir Kumar, two world authorities on the origin of life and authors of the book *"The Timetree of Life"* (2009), point out that the correct names must be: archaea, eubacteria, and eukaryote as initially proposed by Woese and Fox in 1977.

[22]   Prokaryotic cells have genetic information in a region of the cell cytoplasm called the nucleoid, but a membrane does not surround it. They are small cells (1-10 μ). The term prokaryote tends to disappear to be replaced by that of proto-eukaryote. Eukaryote refers to all those cells with their genetic material in the cell nucleus, delimited by a double lipid layer membrane. They are large and more evolved cells (10-100 μ) *(Chatton, 1950).*

Some researchers suggest that the nucleus appeared in archaea and eubacteria to protect against viral infections, giving rise to eukaryotes and limiting contamination of their cellular genome to some extent. Finally, about 2 billion years ago, a type of symbiotic eubacterium called alpha-proteobacteria, similar to the current intracellular parasite Rickettsia, was transformed by endosymbiosis into an intracellular organelle called mitochondria, generator of energy for respiration and cellular metabolism, with its genetic code (DNA), capable of self-replication *(Lithgow, 2010; Wang, 2015)*.

Moreover, a billion years ago, a cyanobacterium of the genus *Cyanophora*, also by endosymbiosis, gave rise to a type of *plastid*, the chloroplast in algae and plants. Also, with its DNA and potential to self-replicate could carry out photosynthesis *(Fig. 2.8)*. Although oxygenic photosynthetic activity *"Photosynthesis II"* occurred 2.8 billion years ago, the ancestors of cyanobacteria, called pro-cyanobacteria, could perform non-oxygen-producing photosynthesis. *"Photosynthesis I"* about 3.5 billion years ago. *(Cardona, 2016)*. All the oxygen produced was initially captured chemically by the dissolved iron and organic matter; eukaryotic cells lived in water in hypoxic environments and metabolized glucose through fermentation. The Great Oxidation described when the rocks' oxygen deposits became saturated due to the Earth's massive oxygenation. Thus, the oxygen molecules escaped into the hydrosphere and the atmosphere. This progressive phenomenon caused the massive extinction of anaerobic microorganisms 2.4 billion years ago. The species that evolved after the great oxidation were characterized by having aerobic metabolism *(Canfield, 2013)*.

In summary, the tree of life formed as follows: LUCA's ancestors arose as complex self-replicating entities with a genetic code that appeared by natural selection *(Fox, 2010)*. They are *"protobionts and progenotes."* After LUCA's integration, the *Archaea* domain originated as thermophilic cells and the *Bacteria* domain as reducing cells that would evolve to acquire extreme thermophilic capacity. Finally, the *Eukarya* domain was born as an emergent process from the other two domains. Over 2.7 billion years, this last domain gave rise to fungi, plants, and animals, whereas archaebacteria and eubacteria are the superkingdoms with the most negligible evolution. Their current structure is very similar to the one when they appeared.

Furthermore, a remarkable discovery was recently made: the eukaryotic supergroup *Opistoconta* is the common ancestor of fungi and animals; therefore, evolutionarily, fungi are closer to animals than plants *(Torruela, 2015)*. Finally, as a side note, it is essential to mention that, although LUCA likely appeared between 100 to 600 million years after the formation of the Earth *(Pearce, 2018)*, this primeval life became extinct due to the aforementioned «Late Heavy Bombardment Period. » This period occurred over three hundred million years, between 4.1 and 3.8 billion years ago, and this has been called «The frustration of the origin of life by impacts» *(Nemchin, 2008; Pross, 2013)*. This theory is supported because the oldest fossils found in March 2017 in Nuvvuagittuq, Canada, were 3.77 billion years old. These relics are hematite tubes embedded in quartz crystal formed around hydrothermal vents in the early ocean. In addition, the 3.7-billion-year-old stromatolites found in Isua, Greenland, are the oldest known life relics *(Dodd, 2017)*.

In April 2016, an international group published an extensive investigation entitled *"A new view of the Tree of Life,"* where they revealed the construction of an organizing principle of unpublished Biology *(Fig. 2.9a)*, with the information of 1,011 genomes of various organisms that came to revolutionize the knowledge that had had for decades *(Hug, 2016)*. This publication and others *(Eme, 2017; Castelle, 2018)* found that the closest phylogenetic relationship of all ancestral eukaryotes was with a group of archaea called *"Asgard."* This group consists of five lines of descent, of which four remain archaea to the present day, except for the fifth, which are the -eukaryotes- whose differences with their three cousins and sister (Fig. 2.9b) lie precisely that they achieved endosymbiosis with an *alpha-proteobacterium* that, as it evolved, became a *mitochondrion* in animals and fungi, and in plants and algae endosymbiosis occurred with descendants of *Cyanobacteria*, which became *chloroplasts*.

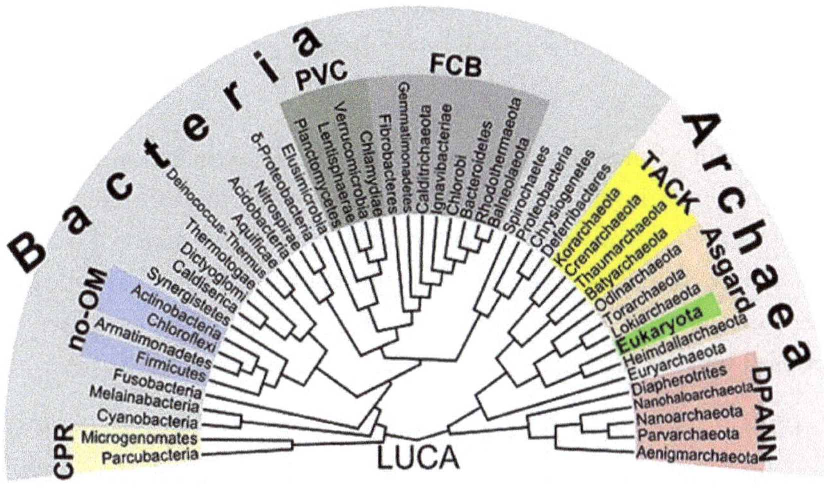

Figure 2.9a   Updated phylogenetic Tree of Life.
A eukaryote is a subclade within Archaea.

Credit: Hug LA. et al. (2016) A new view of the tree of life. *Nat Microbiol.*
1 : 16048. 76. Castelle, C.J: (2018). Cell 172 (6): 1181-1197. (CC BY-SA 4.0).

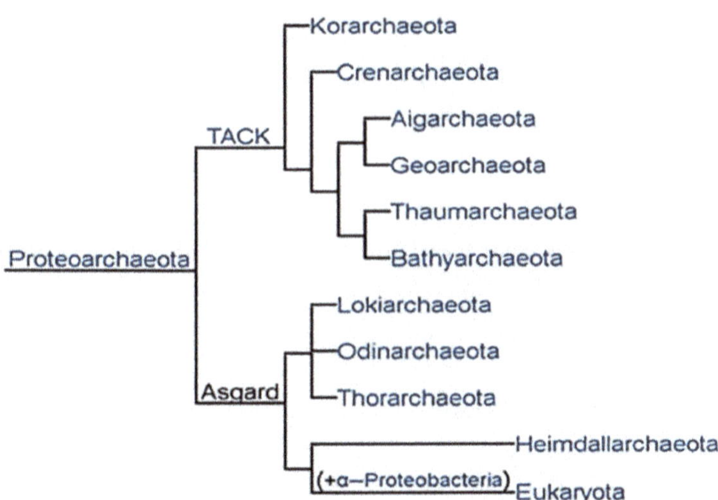

Figure 2.9b   Origin of Eukaryotes.   **Cladogram: Protoarchaea →**
Superphylum Archaea *Asgard* → Eukaryotes (+ α Proteobacteria).
Credit: Veeroja, K. (2020) Models for Eukaryogenesis.
The hydrogen hypothesis. *Bachelor's Thesis.* Tampere University. Finland.

Without this endosymbiosis and cell differentiation, it is unlikely that all multicellular life could have developed *(Eme, 2017; Veeroja, 2020).*

As a result of these findings, it is now concluded that all multicellular organisms have both archaea and bacterial genomes, so the current Phylogenetic Tree has only two direct descendant domains of *LUCA*: *Archaea* and *Bacteria.* —*Eukarya* is a subdivision of Archaea and not the other way around or of simultaneous origin, as the Tree of Life proposed by Woese at the end of the last century, shown in *Fig 2.8.*

On the other hand, now we have a «Catalog of Life» for the hierarchical organization of living organisms that *(Ruggiero, 2015)* returns to the classification proposed by Chaton in 1925 and Allsopp in 1969 with two super-kingdoms that encompass all representatives of Cell Biology: The *Prokaryote* and the *Eukaryote.*

However, it seems that things are not so simple. A recent discovery *(Yamaguchi, 2012; 2014; 2018; 2020)* made more than 1,200 meters (3,900 ft) deep in the -Sea of Japan- locates a unique entity with intermediate characteristics between prokaryotes and eukaryotes, which the discoverers called *Parakaryon myojinensis.* This microorganism has nucleoid and endosymbionts, similar to eukaryotes. The genetic material of DNA is stored in filaments like prokaryotic bacteria and not in linear chromosomes. Furthermore, its cell wall composed of peptide-glucans is similar to bacteria. Ribosomes are in the cytoplasm and the nucleus, and their endosymbionts are some types of bacteria but not mitochondria or chloroplasts. In such a way, it seems that has been found the missing link in the evolution of prokaryotes to eukaryotes forms a new intermediate kingdom called *"Parakaryota."*

As we can see, the first proposed tree of life has expanded dramatically due to new genomic samplings of previously unknown microbial lineages. This description is a global vision of the diversity within each main lineage, and the results reveal the predominance of bacterial diversification over the Archaea domain.

Finally, as a result of this section, this is the hypothetical sequence of events that gave rise to life contained in the magnificent article published in 2008, *"What Is Life? What Was Life? What Will Life Be?,"* and in 2015 *"Hydrothermal conditions and the origin of cellular life"* by Dr. David W. Deamer from the Department of Biomolecular Engineering at the University of California at Santa Cruz *(Deamer, 2008 & 2015).*

This is his literal exposition:

• *The most plausible site for the origin of life was not the open ocean, ice fields, or dry land. Instead, there is reason to think that a freshwater environment would be most conducive to life's beginnings, in the form of an interface between mineral surfaces such as volcanic lava, a body of liquid water, and the early atmosphere.*

—The hypotheses of hot springs (geysers) near volcanoes, and the oceanic hydrothermal vents have implications related to the Panspermia theory. If life originated from deep-sea vents, it is unlikely it did so from the Red Planet. But if life on Earth arose near volcanoes from terrestrial hot springs, it could have started on Mars, which also had the thermal ingredients of widespread volcanism and water *(Van Kranendonk, 2017).*

• *The site's local environment would not be a small hot pool, as Darwin had suggested in a letter to his friend Joseph Hooker in 1871. Instead, it seems more likely that it resembled the types of acidic thermal pools we see today in some volcanic regions, which constantly change through cycles of drought and flood.*

• *The ponds contained complex mixtures of dilute organic compounds of various origins, including extraterrestrial material that arrived at the end of the accretion period and other compounds produced by chemical reactions associated with volcanoes and atmospheric reactions.*

• *The compounds were of two varieties. One variety included those soluble in water that were monomers with the ability to chemically couple into polymers. Examples include amino acids, nucleobases such as adenine and uracil, and simple sugars such as glyceraldehydes, ribose, and phosphate. The second variety comprised amphiphilic compounds such as fatty acids and fatty alcohols that self-assembled to form membranous compartments.*

• During the drying cycle, the diluted mixtures are concentrated into thin films on the mineral surfaces; this concentration process is necessary for the chemical reactions. Under these conditions, the compounds do not simply react with each other, but the products of the reactions are encapsulated within the microscopic compartments produced by the amphiphilic compounds.

• The result of this process was that large numbers of protocells which appeared everywhere on early Earth, where water solutions had cycles of humidity and drought in volcanic environments similar to those of Hawaii or Iceland today. Protocells are defined as compartmentalized systems of molecules, each different in composition from one another. Each represents a kind of natural experiment.

• Most of the protocells remained inert, but some may have had content that could drive them toward greater complexity through energy capture and fewer molecules outside their encapsulated components. The smaller molecules were transported into the internal compartment and required energy to bind them together in long chains. The nature of the reactions that underlies this process represents a significant gap in our knowledge of the origin of life.

• So far, most of the processes that led to the production of protocells fall within the realm of physics, so in a genuine sense, physics came first on the road of life. However, later, through the laws of physics, the first chemical reactions could occur in microscopic vessels. These reactions began the evolution towards culture, the dividing cells that were precursors of what we now call bacteria.

• The smallest molecules are monomers, and the long chains are polymers. Monomers called amino acids form polymers called proteins, and nucleotides form nucleic acids (DNA and RNA). Biopolymers have emergent properties beyond what monomers can do on their own. Most importantly, both types of biopolymers today can act as catalysts. One of them, nucleic acids, can carry and transmit genetic information through the specific sequences of the monomers that compose it.

• Life began when a small fraction of the vast numbers of protocells found a way to grow and incorporate a cycle involving biopolymers with properties related to

catalytic function and genetic information. Thus, cells, not molecules, were the first forms of life. The passage from chemical protocells to simple cells with minimal life properties is a significant gap in our understanding of the origin of life.

Finally, concerning the final part of the title of the 2008 article, what will life be? Dr. Deamer mentions that a significant amount will be artificial life, different from a simple copy of life as we know it. Instead, it will represent a second origin, a version or versions with intelligent designs, elaborated by diverse scientific groups supported by artificial intelligence, and finally achieve the breakthrough.

Synthetic biology is already in full swing with creating biomolecules and engineering biological systems with new programmable functions not found in nature. The Synthetic Genome Team of the J. Craig Venter Institute has already created a new bacterium called *Mycoplasma laboratorium (Gibson, 2010).*

Although Synthetic Biology projects are mainly directed toward Biomedicine for drug development, cell reprogramming, gene therapy, tissue repair, and regeneration. However, it is not difficult to imagine the dire consequences if these investigations are used for military purposes.

From Unicellular to Multicellular Life and the Rise of Cancer

After the emergence of single-celled life on Earth, the next big step was the evolution of pluricellular and multicellular life. *Multicellularity* is characterized by the collaboration between cells of different types that will constitute tissues and organs for a particular organism's development, maintenance, and reproduction. The multicellular organisms have permanent cell junctions, i.e., cells have lost their ability to live independently, requiring association. On the other hand, *Pluricellularity* implies several cells of a single type. Pluricellular organisms result from the union of unicellular individuals through forming colonies, filaments, or aggregation.

The Theory of Cooperation derives from the underlying processes that favor the transition from unicellular to multicellular organisms. This multicellularity evolves due to the formation of groups of cells with new physiological and behavioral capacities that give rise to some advantages concerning unicellular life forms, allowing the possibility of combining greater complexity with a better chance of survival *(Aktipis, 2015)*. However, multicellularity must have strict rules to be successful and achieve a harmonious and collaborative coexistence.

Evolution's first step to achieving multicellular entities was to form colonies of interacting cells and communicate between them through chemical signaling. Today, we can observe these first attempts at multicellularity-pluricellularity with a protozoan, an amoeba called *Dictyostelium discoideum*. This amoeba that lives in the soil and the fallen leaves in the forests is known as "soil mold." It has unicellular life as long as it has enough food sources, the bacteria. Still, when they are scarce or completely absent, it begins to send chemical signals in the form of pulses that attract other amoebae, and this process is called chemotaxis. Then many cell units join together, and the larger the conglomerate, the greater the chemotaxis to the extent that tens of thousands of them participate in about 24 hours. Once united, they start cooperating like a multicellular organism and move in unison. Although they never really fuse, they constitute a pseudoplasmodium similar to garden slugs, about 2 to 4 mm long. It can move at speed impossible for each cell separately.

This pseudoplasmodium or mobile slug undergoes a second transformation. It enters a sporulation program where the cells of the anterior part of the slug form pre-stems of multiple shapes that progressively become true stems at whose tips will form the fruiting bodies of the rear part of the slug. Finally, the cells that make up the stems will die.

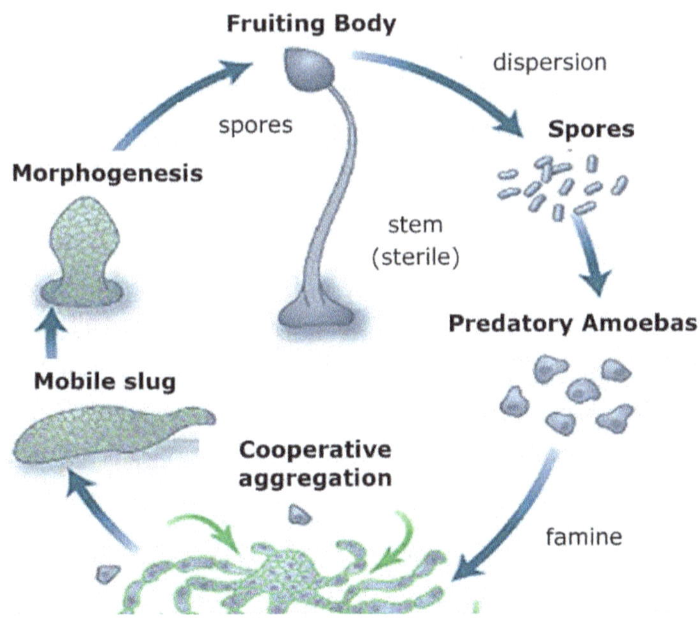

Figure 2.10 Adaptive cycle of Dictyostelium discoideum.
Modified from: Crespi, B. & Springer, S. (2003).

Those that make up the fruiting bodies will transform into spores that will disperse throughout the environment, surviving for extended periods even under extreme conditions. Then, when there are favorable environmental conditions, they germinate and transform into amoebae and thus start a new cycle depending on the abundance of food sources *(Fig. 2.10)*.

However, there is still more to be amazed at the behavior of these microorganisms. For example, it was recently discovered that a third of these spores carry live bacteria. When these spores reach their new habitat, the bacteria are released to reproduce and form the colonies that will serve as food later when they germinate and transform into amoebas again. This activity is considered a type of sowing and harvesting, that is, primitive agriculture. Thus, *Dictyostelium discoideum* is a social amoeba representing alternative unicellular-multicellular life *(Crespi, 2003)*.

While multicellular life arose from single-celled life entities, the latter remains the dominant life form on the planet that often exists in a symbiotic or parasitic relationship with various multicellular life forms. The nature of such relationships can significantly influence the life cycle of plants and animals. An example of this is the development of the complex microbiota in mammals that are essential in acquiring simple and sophisticated functions and constructing and developing all organs. For instance, in the case of the brain, it has recently been identified that the gut microbiota has played a critical role, specifically in cognitive function and basic behavior patterns, such as those that facilitate social interaction and regulate environmental stress *(Dinan, 2015)*. That is the so-called Microbiota-Gut-Brain Axis.

However, estimates made between 2005 and 2010 about the relationship between the number of bacteria that are part of the total microbiota in humans and their number of cells indicated that it was 10:1, that is, ten bacteria for each human cell. Nevertheless, accurate calculations based on systematic analysis suggest that this relationship is wrong, and the correlation is very close to 1:1. Indeed, the number of bacteria is ~ 38 billion —with a total weight of about 200 g. This number is practically equal to the number of cells. Furthermore, in his book *"Genetics of the evolutionary process,"* the Russian geneticist Theodosius Dobzhansky, " published in 1970, pointed out that humans are made up of about 7×10^{27} atoms (7,000 quadrillion atoms). Today we know that they are grouped within 30 to 37 billion cells that make up the adult human body *(Bianconi, 2013; Sender, 2016).*

Multicellularity represents a set of innovations for cellular cooperation to exist. Still, this multicellularity also offers opportunities for cells to "cheat," exploiting the infrastructure and resources of the body they make up. Furthermore, multicellular beings evolve less quickly than the cells that make them up, unleashing vulnerability to cellular innovations that can cause cancer. Multicellular bodies implement additional innovations to counter these threats, including the adaptive immune system and partnerships with specific cooperative organisms.

Thus, multicellular organisms maintain control in the internal fight between deception and cellular cooperation to preserve life *(Aktipis, 2017).*

The evolutionary bases of multicellularity are based on the five elements that constitute the five elements of the *Theory of Cooperation:*

1. Inhibition of cell proliferation.
2. Regulation of cell death
3. The division of labor.
4. The transportation of resources.
5. The creation and maintenance of an extracellular environment

1. Control over the proliferation of cells is necessary for efficient multicellularity to exist that allows the progress, conservation, and proper functioning of tissues. For example, the body of a mammal is made up of more than 200 different cells *(Schneider, 2014);* however,

except for stem cells and their descendants and a few highly regenerative tissues such as the liver, the capacity for proliferation is strictly repressed.

2. To regulate proliferation, multicellular organisms have evolved redundant controls in the cell cycle and mechanisms that automatically trigger apoptosis or programmed cell death and senescence. In this way, these systems suppress biological traps that favor uncontrolled cell proliferation. Conversely, the lack of inhibition of proliferation and regulation of cell death leads to uncontrolled cell replication, which is one of the central characteristics of cancer.

3. On the other hand, effective multicellularity requires integrating cells that make up various tissues capable of performing specific functions, thus allowing labor division. The division of labor evolves spontaneously in multicellularity. Therefore, it is vital to control the differentiation process generated by the different types of cells during balanced multicellularity to preserve the tissues which will form effectively. Inappropriate differentiation of tissues is considered a central feature of cancerous tumors.

4. All cells require resources to survive and carry out their functions. Multicellular organisms typically have a more efficient and less wasteful metabolism than single-celled organisms. Still, large multicellular aggregations require resource transport systems, i.e., conductive vessels, because cells in tissues cannot meet their needs, such as oxygen and nutrients, only through diffusion. Disruption or manipulation of resource transport systems is a central feature of cancer, such as angiogenesis (induction of new blood vessel growth) and metabolism deregulation.

5. Multicellularity also requires the equitable allocation of resources, the proper functioning of cells, and creating and maintaining a shared environment. Waste produced by functioning cells in a multicellular body must be cleaned up, and dead cells must be identified

appropriately and recycled. Similarly, cells need to maintain their extracellular matrix, which comprises protein networks that form supporting structures like basement membranes. Cancer cells destroy the extracellular matrix using various factors, for example, metalloproteinases, thus facilitating their invasion *(Aktipis, 2015)*.

The nature of cancer is inherent to multicellular life; it affects plants, fungi, and animals with greater or lesser frequency among different genera and species. Some life forms have not been observed, such as marine sponges, jellyfish, flagellates, certain worms, and placozoans, originating before the Cambrian explosion. Still, interestingly, most people think that the disease is exclusive to humans.

Genetics undoubtedly plays a fundamental role in the origin and nature of cancer. From the beginning of cellular history, many genes have been exchanged by horizontal transfer, a process in which an organism or cell transfers genetic material to another cell that does not include its descendant.

For example, viruses invaded cells of early mammals, and viral genetic relics became embedded in their junk genes (genes that do not code for proteins). However, not all junk genes are genetic relics of viruses.

These viral relics have been inactivated throughout evolution and have little impact on the lives of animals. However, evidence suggests that some of these are linked to cancer. Cancer develops when cell reproduction is impaired by environmental factors and due to the participation of these viral genes in a process similar to genetic engineering. Thus, cancer cells are cell clusters that arise in genetically modified organisms.

There are two main cell groups in the development of cancer. First, a group of cells appears by natural genetic engineering from a viral genetic relic, such as the endogenous retroviruses that developed after the oxygenation of the atmosphere 2.4 billion years ago. Viral genes genetically modify the organisms with this group of cells. These cancer cells can be inhibited with various antineoplastic treatments. The second group is older, 2.7 billion years ago, and also arises by biological engineering of the genes of ancient eukaryotic cells that existed before the oxygenation of the Earth. This second group in which these cells are present is known as genetically modified organisms by eukaryotic genes. Initially, this group of cells lived in hypoxic environments and metabolized glucose by fermentation. Its genes are included in the genome of various animals, including man, and represent advanced cancer as they increase more aggressively and are resistant to DNA damage and destruction, even with combined cancer therapies.

It is known that as an endogenous retrovirus becomes more abundant in the genome of a mammal, the chance of that mammal developing cancer increases. The hypothesis also affirms that most cancers originate precisely in those organisms genetically modified by viral genes.

However, as cancer progresses, other cancer subgroups will develop in organisms modified by eukaryotic genes, thus giving rise to two cancerous cell lines. Furthermore, because the genes of ancient eukaryotes improved innate immunity, they finally prevailed over the weakest, the viral genes. In this way, during the final stages of cancer, the most resistant and aggressive cells predominate *(Chan, 2015)*. Thus, cancer eventually collapses the cooperative cellular system.

The genome of normal cells is in a dynamic balance between the processes that generate mutations and the processes that maintain the correct sequence of nucleotides. Repeatedly, many reactive cellular metabolites and specific environmental agents alter the chemical structure of DNA, and these alterations lead to mutations in subsequent replications. Even DNA polymerases can generate mutations even when they are copying undamaged DNA.

Fortunately, many repair processes restore the normal nucleotide sequence in the opposite direction. As a result, in normal cells, base sequence duplication in nuclear DNA is exceptionally accurate, with a rate of only one error per billion polymerized nucleotides *(Loeb, 2003)*.

In 1914 the German embryologist Theodore Boveri (1862-1915) published the first scientific article about the causes of cancer; entitled: *"About the origin of malignant tumors,"* where he suggests that the loss of cellular attributes now known as "tumor suppressor genes" are the key to cancer development and that heredity may play a role in susceptibility to cancer himself *(Hansford, 2014)*.

Since the middle of the 20th century, the relationship between cellular genetic mutations and the appearance of cancer has been recognized *(Nordling, 1953)*. In 1960 the first abnormal chromosome related to cancer was discovered; the *"Philadelphia Chromosome"* present in patients with -chronic myeloid leukemia disease- due to a translocation[23] between chromosomes 9 and 22 *(Nowell, 1960)*.

In 1971 Alfred G. Knudson (1922-2016) published his *"double hit"* hypothesis for the development of retinoblastoma in humans. His theory says that the first hit is family inheritance due to the mutation of an allele gene. The second hit occurs due to various intrinsic and environmental causes that cause the mutation of the second allele gene, leading to cancer cells' appearance *(Knudson, 1971)*.

In 2011, it was discovered that certain cancers could appear due to a catastrophic "single" break of any chromosome into several pieces during mitosis, with subsequent multiple abnormal rearrangements. This event called -*chromothripsis*- can occur for unknown causes, but mainly due to exposure to ionizing radiation *(Stephens, 2011)*, and this mechanism is essential for the development of bone cancers.

In 1976, a single avian sarcoma virus gene was found to transform DNA from normal genes into tumor genes in certain birds *(Stehelin, 1976)*. Later, in 1979 Opperman published that this avian gene *(src)* is present in the DNA of some human beings despite not having been infected. That and subsequent releases confirm the theory of cancer in organisms genetically modified by viral genes and eukaryotic genes

---

[23] In Genetics, a chromosomal translocation is the displacement of a chromosome segment to a new place in the genome.

(Oppermann, 1979). In 2005, the *"Cancer Genome Atlas Project"* began to catalog genetic mutations responsible for cancer using genome sequencing and bioinformatics; in this way, we improve our understanding and ability to diagnose, treat and prevent *cancer (NCI, 2019).*

As a result, we can conclude that the genome of the affected organism initially determines the development of cancer, to which physical, chemical, or biological environmental factors can be added, and these environmental factors can be producers, triggers, or facilitators. Cancer as a whole is the most complicated cellular aberration with which medical science has been confronted. *(Messerschmidt, 2017).*

The World of Viruses, the Great Kingdom of Life Ignored

Molecular Biology (BM) is a science that falls between the fields of Biology, Genetics, and Biochemistry. It deals with the structure, biosynthesis, and functions of DNA and RNA at the molecular level and examines their interactions with proteins. The central dogma of the BM states: *"DNA is transcribed into RNA and translated into a Protein" (Crick, 1957).*

On the other hand, in simple terms, a virus is a macromolecular complex made up of a generally small DNA or RNA genome protected by a capsid and, in some cases, by an envelope obtained from a previously infected cell. They are called virions when they contain this envelope of lipoproteins or glycoproteins. The virion is the complete functional virus capable of infecting living tissues. All viruses are intracellular parasites and reproduce through the cell's genetic machinery they infect. Unlike bacteria, discovered by Leeuwenhoek in 1676, viruses, due to their size, had to wait for observation until the 1930s, thanks to the invention of the Electron Microscope by Ernst Ruska and Max Knoll between 1931 and 1933 *(Smith, 2018).*

At that time, the Cell Theory was already predominant, strengthening daily: *"the cell is the fundamental unit of life."* In scientific thought, life was (and for some, still is) exclusively linked to cells; thus, everything that was not made up of cells could not be considered alive.

However, no phrase is better in the evolution of human thought than one that says that *"the only permanent thing is change."* We know there is no universal definition of life, only concepts of "cellular life" based on their Theory *(Zhuravlev, 2006)*. We must assume that *«the transition to life is a continuum» (Luisi, 1998)*. In that case, the main task is to investigate and reconstruct a series of transitional forms between highly complex (almost living) supramolecular prebiotic chemical systems and early biotic primitive organisms (already alive). In other words, this point is formulated as the "minimum life" problem. When we can find a clear qualitative boundary between non-living and minimally living systems, the definition of life problems will be immediately solved. Unfortunately, that is what we cannot do *(Tsokolov, 2008)*.

The Last Universal Common Ancestor (LUCA) probably represents a bacteria-like community of protocells bounded by minerals rather than lipid membranes. LUCA was already a living being *(Cornish, 2017)*, but it was not the first life form on Earth *(Fox, 2010)*. Instead, a proto-ribosome probably preceded LUCA *(Fig. 2.11)*, and some authors consider it the link between abiotic systems and the first vestiges of life *(Sleep, 2018)*. Thus, as its name indicates, the acronym LUCA should be the last universal common ancestor of "all life." Still, traditionally it has only been of cellular life, so its correct name should be LCCA (Last cellular common ancestor).

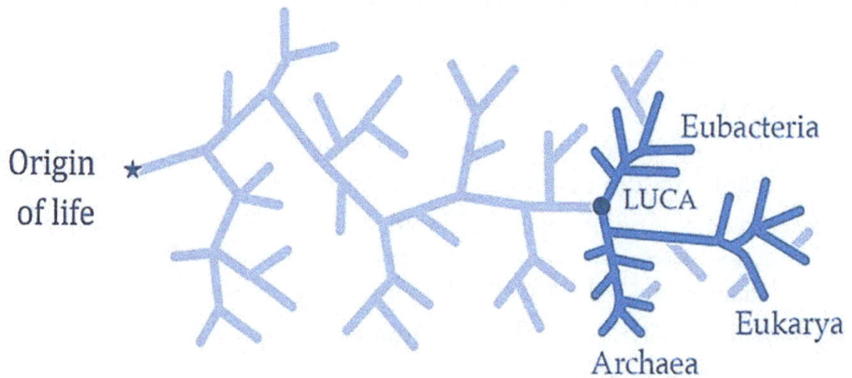

Figure 2.11 The origin of life and LUCA.
Many extinct living organisms were ancestors of LUCA.

Modified from: Cornish, B. (2017). Life before LUCA.
J Theor Biol. 2017 Dec 7; 434: 68-74.

Some think that at some point, an RNA protocell "invented" DNA to store genetic information and also proteins to carry out catalysis with greater efficiency *(Wieczorek, 2012; Torres, 2016)*, a phenomenon that is difficult to explain without the intervention of at least in part of the viruses. However, LUCA was a virus-host, and there is ample evidence of viral evolution before LUCA with the kingdom *Riboviria*. RNA viruses were the first to emerge, but by the time LUCA appeared, they had already been largely replaced by a more efficient DNA-based virosphere such as the *Duplodnaviria* and *Varidnaviria* kingdoms *(Krupovic, 2020)*.

Viruses are undoubtedly ancient and could precede cells or appear at the same time because it is very likely that the "invention" of DNA happened through the —reverse transcriptase— of the ancestors of retroviruses, the *«retrotransposons»* *(Forterre, 2010; Trifonov, 2016)*. Although many eukaryotic cells today possess this enzyme, it was probably due to the ancient infection of cells by these viruses that integrated their genome into germ cells as an aggregated cellular gene. They will continue in the genetic material of the descendants; that is the reason around 8% of the human genome are endogenous retroviruses *(López & Fasching, 2015)*, and not only that, but they are a reservoir from which new infectious retroviruses can arise through recombination *(Heidmann, 2009)*.

Endogenous retroviruses can affect the expression of other genes and therefore influence our development. It is believed that they have influenced our evolution as humans, affecting cell differentiation throughout our embryonic development. They may be related to certain genetic diseases and some types of cancer. Nevertheless, some recent reports indicate reverse transcriptase originated early in the RNA world. Ribozyme RNA polymerase already shows reverse transcriptase and integrase activities. It plays a vital role in the origin of life of all living organisms, cellular and acellular, thus creating the DNA world *(Mustafin, 2019)*. In this new evidence, the infection of the first retroviruses to cells occurred in the sea 460 to 550 million years ago *(Hayward, 2017)*.

Now, it is a fact that viruses are the dominant biomolecular entities on planet Earth and show a genetic diversity far superior to that exhibited by all known cellular life forms *(Ortega, 2020)*. They parasitize all forms of cellular life. Together with the subviral agents or particles called "selfish," they form a world unified by a small set of characteristic genes which encode essential functions in a wide variety of them. In addition, viruses use all the theoretically possible genome replication and expression strategies, in contrast to the uniform replication-expression cycle of cell life forms. The interaction between viruses and the cells they parasitize has been described as a billion-year weapons race in which hosts constantly develop new means of defense.

In contrast, viruses constantly evolve to evade them *(Forterre, 2009)*. The arms race is only one aspect of virus-host co-evolution with all its complexity and essential biological consequences. The other side of coevolution involves cooperation, whereby viruses contribute to cellular functions, while viruses pick up cellular genes and use them for counter-defense and other functions. As a result, viruses create new genetic information, which can be transferred to cells, enhancing biodiversity in all possible life niches. Thus, although the world of viruses maintains its distinction from the universe of cellular organisms during evolution, the genetic exchange between both parts of the biosphere has always been extensive *(Koonin, 2013; Nasir, 2014)*. Regarding their origins, the virus-first hypothesis suggests that viruses predate cellular life forms, appearing in a primitive RNA world from RNA replicons[24] that preceded cellular structures.

Furthermore, this hypothesis suggests that viruses played an essential role in the origin of cells and their evolution through billions of years of parasite-host interaction *(Figs. 2.12a and 2.12b)*. Within the framework of this hypothesis, viroids would be survivors of the RNA world *(Diener, 2016)*.

---

[24] A replicon is a DNA or RNA region or molecule that replicates from a single point of origin.

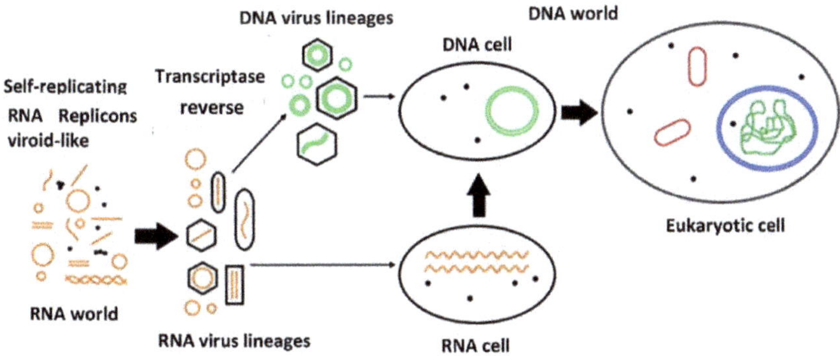

Figure 2.12a Schematic representation of the precellular and coevolution hypotheses.

Modified from: Ortega C.S. (2020). The Origin of Viruses.

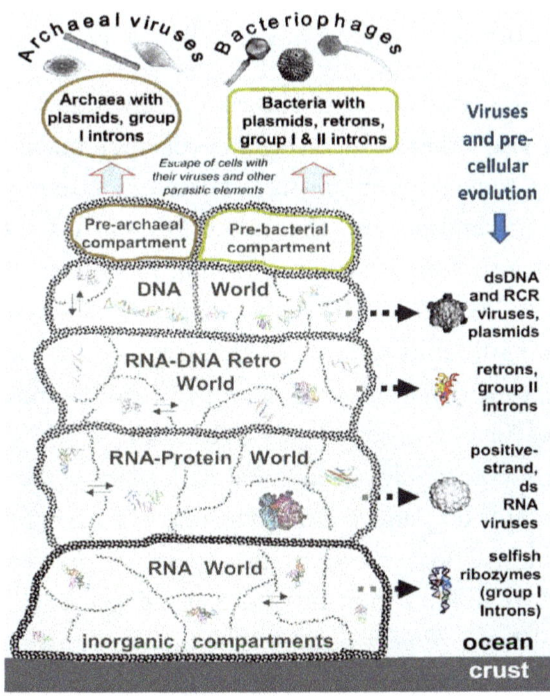

Figure 2.12b Evolution of viruses and cells from the RNA world.

Modified from: Koonin (2006). The ancient Virus World and evolution of cells.

The difficulty with this hypothesis is that all virus species are strict genetic parasites of cells, and it is contradictory to say that viruses pre-date cells. However, this would not be entirely true if they were coevolutionary, had individual or relatively simultaneous origins, or were not obligate parasites at the beginning.

However, the main reason for this section is not to decipher chronologically the origin of viruses concerning LUCA (a situation that, by the way, today has not been solved with certainty) but to broaden the pigeonholed but dominant view of Life-based on cells (Cytota) and understand the reasons why, precisely viruses, should be considered living beings belonging to another domain of life, a non-cellular one (Acytota) and that, both kingdoms together, form the totality of life on Earth or "Biota" *(Fig. 2.13)*. This proposal is very controversial, and it seems that today the history of the second half of the 19th century is repeating itself with the bitter opposition between supporters of the Vital Force Theory against those of the Cell Theory.

So, to start the contest, we could ask ourselves an absurd question: Could viruses and other subviral agents together make up a Tree of non-life? This question is closely related to the work of the International Committee on Taxonomy of Viruses (ICTV), which classifies them into four domains and nine kingdoms with a total of 6,590 species *(ICTV, 2020)*. The abundance of DNA viruses on the planet is around 10^{31} units *(Minot, 2011)*. In fact, to this day, in the study of the viral genome or virome, we can only identify a minor fraction of viral sequences in the order of 10%. The rest of the viral DNA is unknown and is commonly called "viral dark matter." So how are such abundance, complexity, and continuous adaptive evolution possible in non-living beings? Today we know that there are between 10 and 100 viruses per cell of all life forms in the oceans *(Zinger, 2012)*, and on land, the relation is 10 to 1 *(Pratama, 2018)*. The human intestinal virome, one of the microbiota components, is made up mainly of bacteriophages, and their number exceeds the bacteriome by a relation of at least 10 to 1 *(Shkoporov, 2019)*.

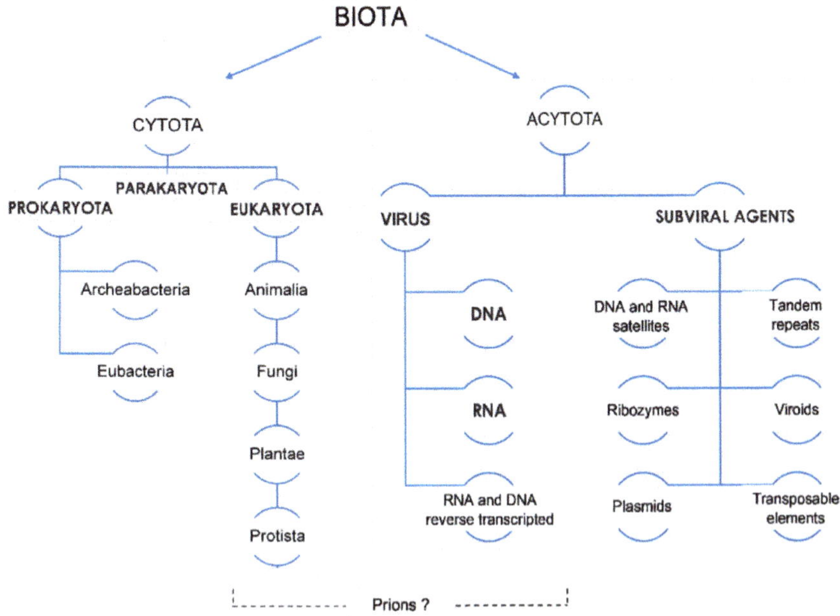

*Figure 2.13   **Hypothetical scheme of the members of Life or Biota.***

Biota is divided into two superdomains:
the cellular or Cytota and the acellular or Acytota.
The cellular has two domains, the Prokaryotic and the Eukaryotic.
In 2012, a third intermediate domain called Parakaryota was found.
The Acytota taxonomic group comprises two subgroups: viruses and the widely
interrelated subviral agents or particles.
As infectious agents made up of proteins, the Prions belong exclusively to non-cellular life.
However, they are encoded from genes exclusive to cellular life.

Credit: Own original work —from Yamaguchi, 2012; Harbi & Ma, 2014;
Trifonov & Diener, 2016; Krupovic, 2020.

On the other hand, the simple distinction based on the size between viruses and cells has disappeared since the discovery in 2003 of giant viruses (Mimivirus, Pandoravirus, Megavirus, Pithovirus), and this has overturned a part of the original definition of a virus. These viruses are larger than many *bacteria (Claverie, 2016);* proportionally, these giant viruses' genome size and complexity overlap the genomic ranges of cellular life forms and exceed those of numerous parasitic bacteria.

As if that were not enough, in 2008, another discovery was made that left the community of biologists speechless: viruses called virophages parasitized giant viruses. Unlike satellite viruses that produce cell co-infection associated with a helper virus, such as hepatitis D virus (satellite) with hepatitis B virus (helper), virophages are DNA viruses that actually infect other viruses. This characteristic has also challenged the definition of viruses because they were only considered cellular parasites. They parasitize the viral factory of giant viruses by hijacking their host viruses' transcription and replication machinery to express and replicate their genomes.

Furthermore, some virophages are authentic parasites that negatively interfere with the host-virus, inducing a partial inhibition of the giant virus reproduction and possibly increasing defective particles' genesis *(Mougari, 2019)*. The discovery of giant viruses and virophages gave a solid push for including viruses in the realm of Biology. However, the world of viruses, or the Acytota kingdom (domain), is vast and comprises diverse molecular entities of all constitutions, sizes, and shapes. They include satellite DNA and RNA, tandem repeats, ribozymes, viroids, plasmids, transposable elements, and viruses. This hierarchy contains everything considered Acytota and suggests kingdom groups on their evolutionary path towards greater complexity *(Trifonov, 2016)*.

On the other hand, "prions" *(proteinaceous infectious particles)* are infectious protein agents that cause neurodegenerative diseases called "transmissible spongiform encephalopathies" that affect some birds and mammals (e.g., mad cow disease), including humans (e.g., Creutzfeldt-Jakob disease).

They can self-replicate in the absence of nucleic acids, and they do so by turning their regular counterpart into a likeness of themselves. A prion is a glycoprotein whose secondary and tertiary structure is altered. The folding or way in which it coils in space is abnormal. In turn, the natural prion protein (PrPc) found in the membranes of cells becomes a pathogenic protein (PrPSc) that, when in contact with Normal proteins, acts as a catalyst for them to transform into misfolded proteins.

Hydrophobic interaction plays a crucial role in standard protein folding, as are hydrogen bonds for RNA folding, leading to base pairing within the chain *(Ma, 2016)*. When these misfolded prion proteins form an extracellular aggregate within the central nervous system, they form plaques known as -amyloids- associated with tissue damage and cell death.

Prions are almost always acquired by ingesting contaminated food (especially the brain and spinal cord). After intestinal absorption, they reach the spleen and lymph nodes from where they travel to the brain. Its latency time is very long and can be up to 5 years. Thus, they are highly resistant to their destruction by physical, chemical, and biological means. The formation of antibodies in response is ineffective in destroying the protein, as are proteases and some organic solvents, detergents, ionizing radiation, or very high temperatures. For example, there is evidence of contaminated materials subjected to a temperature of 600 °C, and they still conserve traces of infectivity *(Cacace, 2011)*. However, their infectivity can be reduced with 10% sodium hypochlorite, 90% phenol, ether, and acetone, and an autoclave for 30 min at 134 °C.

Prionic proteins have a family consisting of prions, quasiprions, and prionoids *(Harbi, 2014)*. Prionoids (such as tau phosphoprotein, α-synuclein, and huntingtin) are related to specific diseases that have been called "prion-like," including Alzheimer's, Parkinson's, Huntington's disease, and amyotrophic lateral sclerosis *(Ma, 2014; Wells, 2019)*.

Recombinant, synthetic prion proteins have been produced in the laboratory and are also infectious for some animal species, and not only that, but like natural prions, they carry out a Darwinian process: «they can adapt to specific environments, predominating the most aggressive prion strain, which translates into the survival of the fittest" *(Legname,*

2017). On the other hand, quasiprions are proteins that exhibit a behavior similar to prions but are not infectious or transmissible between individuals or cells. An example is the quasiprionic mitochondrial antiviral signaling protein, which functions in the innate immune response in humans.

Now, returning with viruses, in 1971, David Baltimore, 1975 Nobel Prize in Physiology and Medicine, made the *«virus classification, »* which, together with ICTV nomenclature, is currently used worldwide *(Baltimore, 1971)*. Viruses are divided into seven groups or classes *(ParaMicroBio, 2019)*, three DNA, and four RNA *(Fig. 2.14)*. Each group is characterized by how it generates mRNA from the virus genome.

Group I - DNA virus ds (double-stranded DNA): The replication of the virus, the transcription to mRNA (messenger), and the translation to the viral proteins are carried out precisely as in the cell, e.g., Adenovirus, Poxvirus, Herpesvirus.

Group II - ss (+) DNA virus (positive single-stranded DNA). The viral genome is converted into ds DNA using the cell's enzymatic machinery and generates the mRNA that will give rise to the different viral proteins. The duplication of the viral genome inserted into the cell is similar to class I. Subsequently, the ds DNA strands are separated before encapsulating the ss (+) DNA virus, e.g., Parvovirus, Chicken anemia.

Group III - ds RNA virus (double-stranded RNA): The RNA strand (+) is used for the transcription, and the formation of the viral genome to encapsulate it is made up of two steps: The first step consists of assembling the ss RNA to the virus (+), and once inside the virus the complementary strand is generated to form ds RNA, *e.g., Reovirus.*

Group IV - RNA ss (+) virus (single-stranded RNA in the positive sense): Since the viral genome ss (+) RNA has the same polarity as mRNA, it does not need any conversion to generate viral proteins.

RNA ss (-) are developed to regulate gene expression and serve as a template to replicate the RNA ss (+). There are two models in this class, the IVa, and the IVb, *e.g., Picornavirus, Coronavirus.*

Group V - RNA ss (-) virus (single-stranded RNA in the negative sense). The virus carries an enzyme called reverse transcriptase (RT) that forms mRNA (ss + RNA), and then it is translated into viral proteins. In addition, viral regulation proteins are also synthesized that generate ss (-) RNA from ss (+) RNA inserted into the virus, *e.g., Rabdovirus and Orthomyxovirus.*

Group VI - RNA ss-RT virus (single-stranded RNA with intermediate DNA in its life cycle). The virus provides a reverse transcriptase that converts ss (+) RNA into ds DNA. The mRNA that forms viral proteins such as ss (+) RNA and the reverse transcriptase will enter the virus, *e.g., Retroviruses like HIV.*

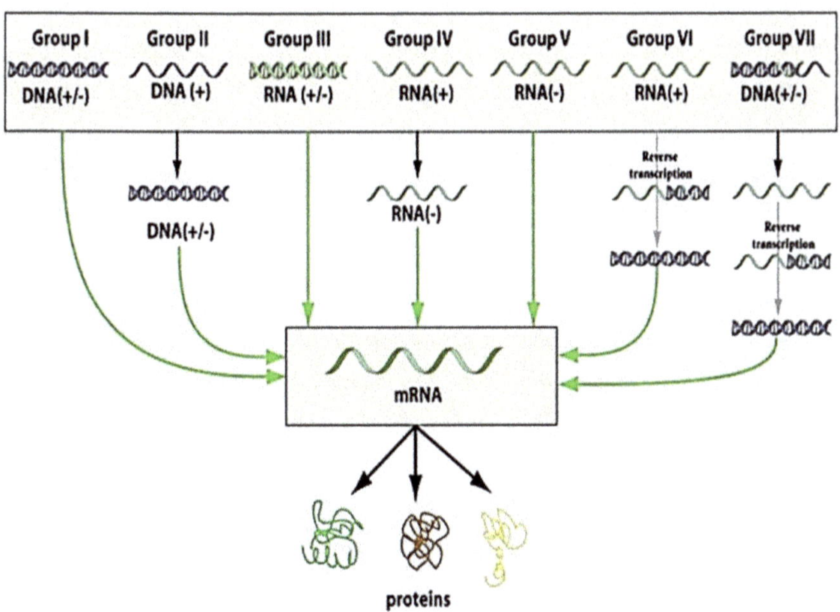

Figure 2.14 *Virus Classification. Baltimore - ICTV.*
Credit: Baltimore, D. (1971) Expression of animal virus genomes.
Bact Rev, 35 (3): 235-241. ParaMicroBio

Group VII - ds DNA viruses (double-stranded DNA with an intermediate RNA in its life cycle): These viruses inside the cells make a transcription of DNA into positive single-stranded RNA and into mRNA, which performs translation for the production of viral proteins, among them, RT which it assembles the virions in an RNA / DNA complex that finally completes their conversion into double-stranded DNA, leaving the cell by budding. *e.g., Hepadnavirus.*

In September 2013, during the V European Congress of Virology in Lyon, France, viruses were objectively defined as living beings; Dennis Bamford, director of the Molecular Virology program in Helsinki, indicated this during his conference "The Order in the Universe of Viruses." None of the two hundred speakers and attendees had a different opinion. Based on this conference, Dr. Miguel Ángel Jiménez Clavero, renowned Spanish virologist and science communicator, published the following comment: *« Viruses are a peculiar type of microorganisms because they are not cells, but subcellular entities. This is the main difference between bacteria and why viruses have traditionally been excluded from the category of "living things." Human beings have defined life in a certain way, which seemed best to us in its day. This definition states that life is made up of cells with their metabolism. Viruses can reproduce in a suitable environment -the cell- putting their metabolism at their replication service. They use genetic material of the same nature as that of a cell, which "understands" the message contained in this material because it is written in the same code as theirs and executes it, directing the synthesis of viral proteins that are made up of the same components than cellular proteins. Viruses have a high organization, and their components have recognizable functions. They can mutate, recombine, and evolve, generating diverse forms to adapt better to the environment. In general, they possess other properties than the rest of the living beings. Therefore, I am convinced that viruses are part of the living world. Whether we call them living beings or not is due to the human need to define concepts, classify, and categorize. Viruses escape from this narrow definition of life that currently prevails. Definitions can and should be revised if this improves the coherence of what is defined. »* (Jiménez, 2013).

In preventive medicine, the concept of life in viruses has been accepted for years. Therefore, viral vaccines made with whole viruses are classified as attenuated live virus vaccines such as Sabin for polio and killed or inactivated viruses such as Salk to prevent the same disease.

In conclusion, viruses are living entities belonging to the *Acytota* taxonomic group. They do not contradict the central dogma of Molecular Biology; on the contrary, they faithfully carry it out using cells to fulfill it. Some even extend it, as with retroviruses that transcribe RNA into DNA. For billions of years, viruses have exerted enormous evolutionary pressure on cellular organisms, leading them to develop molecular and cellular innovations searching for simplicity or complexity *(Nasir, 2014)*. The reasons why viruses were excluded from the concept of life until the first decade of the 21st century *(Moreira, 2009)* nowadays are being left behind as we advance in knowledge and new ideas different from those that traditionally appear. Without a doubt, we are living a historic moment in the conceptual evolution of Biological Sciences.

Artificial life

Ideas about artificial life have their origins in concepts created more than a century ago by Stéphan Leduc, a famous French biologist and professor at the Nantes School of Medicine in the late 19th and early 20th centuries, who contributed to the understanding of the chemical and physical mechanisms of life.

In 1912, Leduc published his most important work, *"La Biologie Synthétique,"* a treatise of sixteen chapters. In his analysis of the possibility of creating artificial life, he mentions: « The synthesis of life, if it ever occurs, will not be the sensational discovery with which we generally associate the idea. On the contrary, if we accept the theory of evolution, then the first dawn in the synthesis of life will produce intermediate forms between the inorganic and organic world. These forms will possess only some rudimentary attributes of life. Other attributes will be slowly added in the course of its development by the evolutionary action of the environment. »

Based on the Cell Theory[25], there is currently a great scientific current experimenting with the synthesis of life, or artificial life. Its surprising results further complicate the elusive concept of life; let us look at some examples.

Dr. Gerald Joyce (whom we already mentioned as the author of NASA's definition of life) and his working group at the Scripps Research Institute in San Diego, California, have synthesized enzymes

---

[25] Cell Theory is the foundation of Biology; It has two components: the first says that all organisms are made up of cells, and the second is that cells always originate from other cells.

from RNA called ribozymes. They are non-protein enzymes and can replicate themselves without the help of any protein or other cellular components. These simple nucleic acids can act as catalysts and continue the process indefinitely, becoming "immortal molecules." There is nothing biological in this system: there are no cells; they are only molecules provided with building blocks, remaining active until they run out of substrate *(Horning, 2016)*. This experiment replicates what probably happened in the early RNA world.

On the other hand, Dr. Jack W. Szostak, a biologist from Harvard Medical School, and his team have researched the construction of an artificial life form in the laboratory with fatty acid molecules in water. These molecules spontaneously form vesicles with membranes similar to biological cells and mimic the replication of DNA molecules. As a result, its protocells can grow and divide and even compete with each other. Moreover, vesicles containing phospholipids in their shells, a crucial component of natural cell membranes, grow faster than those without them *(Jin, 2018)*.

In 2014 Dr. Martin Hanczyc from the University of Southern Denmark published that he, and his workgroup, had created protocells using nitrobenzene oil droplets placed in a highly alkaline solution (pH 12) "fed" with oleic anhydride that turns into oleic acid in contact with water. This reaction lowers the pH at the periphery of the droplets, creating an irregular surface tension that makes them move autonomously through the liquid inside a Petri dish, spinning, merging, dividing, chasing each other, and behaving as if they were alive.

The convection within the droplets provides fresh supplies of oleic anhydride to the surface, and the droplets can "feel" their surroundings moving through a gradient, seeking the highest possible ph. Although they do not contain genetic material, these protocells absorb fuel and react to their environment. Although some critics have mentioned that it is due to Brownian molecular motion, this model is not so since the movement of nitrobenzene droplets does not occur randomly. Still, they seem to have a much more structured pattern called molecular behavior. Only five chemical compounds were used in the experiment, and as Dr. Hanczyc says, this could resemble some tests of early life on Earth *(Hanczyc, 2014)*.

Molecular behavior, although not by that name, was known by alchemists such as Johann Glauber since the seventeenth century with the creation of "philosophical trees" now called "chemical gardens." These gardens are perhaps the best example in the chemistry of a self-organizing disequilibrium process that creates complex structures by using semi-permeable membranes and amorphous and polycrystalline solids. That occurs at the chemistry, fluid dynamics, and materials science interface. This developing field is called *"Chemobrionics" (Fig. 2.15)*.

Figure 2.15   Chemical Garden -formed by adding cobalt, copper, iron, nickel, and zinc salts to a sodium silicate solution (size: 5.5 × 3.7 cm).

Modified from: Barge, L (2015) From Chemical Gardens to Chemobrionics. *Chem Rev* 115: 8652-8703.

Another fascinating area of research related to the creation of artificial life is the synthesis of « recombinant prions » in the laboratory since the end of the nineties, allowing us to observe some incredible facts. These are also infectious like natural prions and cause long-term central nervous system disease in laboratory mice. In 2018, the first pathogenic human artificial prion (rhuPrion - recombinant human Prion) was synthesized, and it also has a behavior similar to its natural counterparts *(Kim, 2018)*. Life-based on proteins such as prions is not accepted even though at the origin of life, the first association of the primitive RNA world was specifically with proteins, forming the *"RNA-protein world"* from which the first cells and viruses emerged. *(Fig. 2.12b)*. These protein elements are replicating without nucleic acids, a phenomenon observed precisely in prions and ribozymes.

As we saw in the previous section, prions have a Darwinian adaptation, survival of the fittest. Therefore, the most aggressive protein predominates when competing with other prion proteins *(Legname, 2017)*.

Concerning artificial cell life, Buddingh and van Hest from the Eindhoven University of Technology in the Netherlands recently published a paper regarding state of the art in the world in the construction of artificial cells. They begin from the fact that one must first consider the minimum criteria for cell life to build synthetic cells.

They use the *«chemoton»* or chemical automaton model, developed by the Hungarian biologist Tibor Gánti and published in 1971 in his book *"The Principles of Life,"* where he describes an entity that comprises:

1) An integrated chemical system.
2) A chemical information system,
3) A self-generating chemical engine" metabolism "that can be considered "alive"; additionally,
4) Growth and reproduction, which are necessary for the survival of the species, and finally,
5) Adaptability is essential for life in diverse environments.

Once defined what is considered a living system, the first thing to be done is the compartmentalization of the future cell to separate its internal environment from the external one using membranes, and then the sub-compartmentalization to simulate the hierarchical architecture of eukaryotic cells in tissues. In addition, synthetic gene circuits must be used to encode the information necessary to create complex behavior such as feedback. Finally, artificial cells must reproduce to maintain their population through controlled growth and keep their synthetic spaces separate; this has already been accomplished, but extensive cell division regulation, as nature does, remains unattainable.

There are several future challenges to overcome; for example, for artificial cells to be autonomous and able to communicate with other synthetic and natural cells, such as bacteria. *(Buddingh, 2017)*.

As we review in Molecular Biology, another essential point to developing artificial life involves precisely separating the living from the non-living. Due to this current around artificial life and the inclusion of viruses within living beings, it seems almost impossible to delimit the concept of life in our world. —To quote a phrase from Friedrich Nietzsche: « There are concepts that can be defined, while others only have history. »

Dr. Carol Cleland, a philosopher of science at the University of Colorado, comments: « I think it is a mistake to define life because we only have one example of life, the known life on Earth, and we have reasons to believe that this example may not be representative of life in general. »

That is why current research on the possible chemical basis of extraterrestrial life is being carried out comprehensively and in detail. Massachusetts Tech Institute astrophysicist Sara Seager and her collaborators have thoroughly analyzed and developed a list of possible chemical combinations that could indicate the presence of extraterrestrial life. Making computer-guided mixtures of the six chemical elements essential for life to exist (CHNOPS), «Carbon, Hydrogen, Nitrogen, Oxygen, Phosphorus, and Sulfur, » including combinations never before seen on our planet. Some of these mixtures are probably unusable, but this is the beginning of the road to identifying alien entities in the future through their biochemical signatures.

Having reviewed all these concepts, we must recognize the extraordinary value of the contributions of so many researchers whose ingenuity has brought us to the level of knowledge we have today, starting with Schrödinger and his ideas on the processing of matter and energy for the generation of life, which, although reductionist, can be generalized to any type of alien existence. That is because they are based on the laws of physics and the cosmological principle. We still lack a precise and complete definition of life and partially ignore its origin. However, in science, we must not forget that the most exciting questions are precisely those that cannot be answered.

Present and Future of Life on Earth

The Terrestrial Biosphere is the global sum of all ecosystems. It is just 28 kilometers thick (17.4 mi), 17 kms (10.5 mi) of them above sea level to the edge of the troposphere or tropopause, *"the thin blue line,"* and ~ 11 km (6.8 mi) below it. The Earth's radius at the equator of 6,378 km (3,963 mi), so the biosphere is approximately 0.4% of this distance. Thus, all life known in the Universe until today is reduced proportionally to the thickness of the peel of an apple taken to the surface of a small rocky planet.

An estimate of the total DNA in our Biosphere has recently been made. It corresponds to 5.3 x 10^{31} megabases of DNA with a rate of 10^{24} nucleotide operations per second, equivalent to the processing capacity of 500 million of the four most powerful current supercomputers in the world *(Landenmark, 2015)*. This gigantic magnitude is how living beings interact with each other and our environment. That is one of the foundations that our entire planet functions as a superorganism that has been called Gaia.

There are conditions for life to exist only during one stage of existence in the Universe. Although original calculations of the famous physicist Robert Henry Dicke in 1961 indicated that the Universe required approximately 10 billion years to make possible the formation of planets with enough carbon to give rise to life. Today, we know that the first stars with carbon atoms in their interior appeared as early as 100 million years after the Big Bang *(Mashian, 2016)*. Therefore, the formation of planets with traces of carbon occurred a few million years later, and life could have started at any time 400 million years after the Big Bang *(Loeb, 2016)*. On Earth, life will undoubtedly cease within a few billion years if we do not self-destruct first or if some galactic phenomenon that is lethal to the planet's biology occurs.

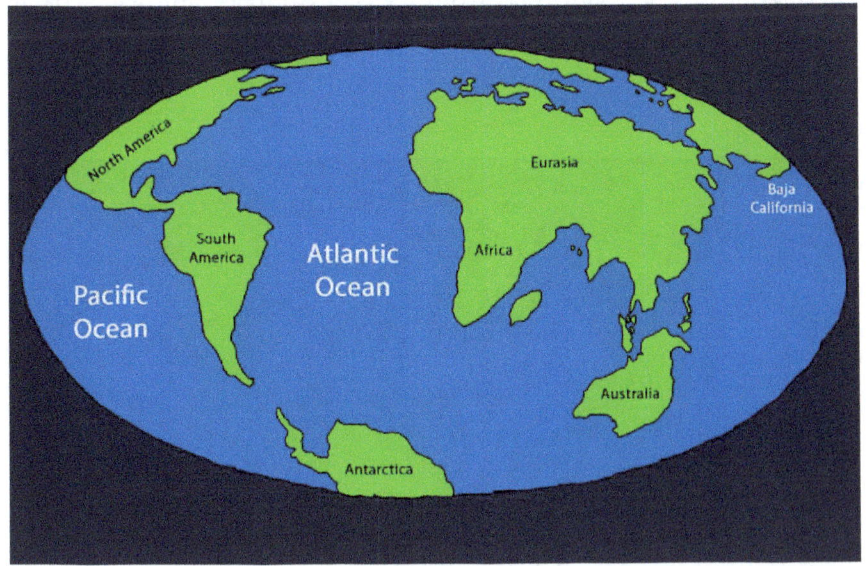

Figure 2.16 (a). Projections of the Earth in 50 million years.

The Atlantic will expand, Africa will collide with Europe closing the Mediterranean,
Australia will collide with Southeast Asia and California,
and Baja California will slide north to Alaska (red arrow).

Credit: Alejandro Betancourt. Image-based on - Christopher R. Scotese, Paleomap.

The Earth, in the future, without any human intervention or extraordinary phenomena, will continue to change as in the past. The climate will become warm, then cold, alternating again and again. Finally, the extensive glaciations will return, and the Earth will perhaps snowball again for the third or fourth time, returning to stages of extreme desert heat.

Tectonic plates will continue to move, open and close oceans, and form a new supercontinent 250 million years from now as Rodinia, Pannotia, and Pangea were before *(Hazen, 2013).*

Figures 2.16 a, b, and *c* show the future of continental drift in 50, 150, and 250 million years. Over the next 50 million years, America will move further away from Europe and Africa, and the Mediterranean Sea will disappear. The Californias from the San Andreas fault will separate from the Continent and migrate north to Alaska.

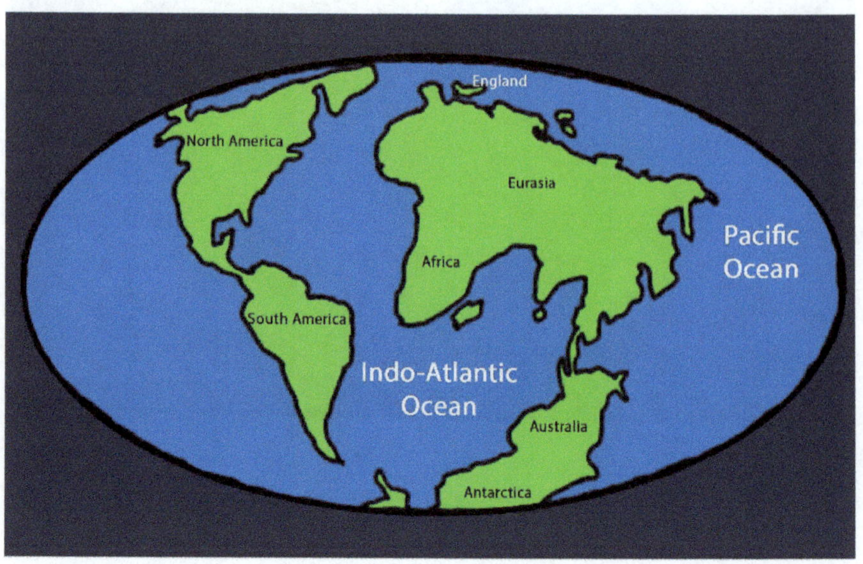

Figure 2.16 (b). Projections of the Earth in 150 million years.

The Atlantic Ocean (then Indo-Atlantic in the south) will begin to close as new subduction zones along the eastern coasts of North and South America will start to consume the ocean floor that currently separates -North America from Africa.

The intercontinental retreat will cease in about 100 million years, and 50 million years later, the continents will have receded enough to halve the intercontinental oceanic space. The Indian Ocean will merge with the southern hemisphere, the Atlantic.

In addition, Antarctica will have migrated to the north and Greenland to the south. As their ice sheets melt, the average level of the seas and oceans will increase by around 90 meters throughout the planet.

Finally, in about 250-350 million years, the *Next Pangea* or *Pangea Proxima* will form, remain for ~ 100 million years, disintegrate again, and constitute new continents.

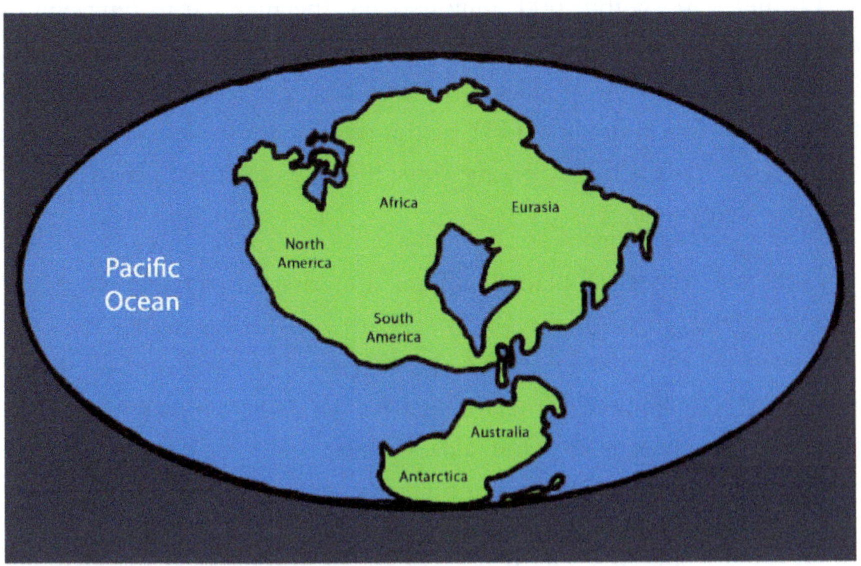

Figure 2.16 (c). Projections of the Earth in 250 million years.

"Pangea Proxima" will form as the subduction of the Atlantic Ocean continues. This supercontinent could have a small ocean basin (a new Mediterranean Sea).

Credit: R. Alejandro Betancourt. Image-based on - Christopher R. Scotese, Paleomap Project (http://www.scotese.com)

Currently, different aspects of the future of the Earth are very detailed, such as climatic changes, geodynamics, glaciations, and solar evolution. Peter Ward and Donald Brownlee's book: *The Life and Death of Planet Earth: How the New Science of Astrobiology Charts the Ultimate Fate of Our World* is an excellent source to understand what tomorrow will happen to our planet *(Ward, 2004).*

Summarized: we can mention that the Earth, in the future, will receive mega impacts from comets and meteorites again, and supervolcanoes will erupt again, causing mass extinctions, but not the disappearance of life *(Hazen, 2013)*.

Furthermore, within 600 million years, the increase in temperature on Earth due to the rise in the sun's luminosity will lead to an increasing weathering of silicate minerals, leading to an increase in the trapping of carbon dioxide. Carbon trapping by the rocks will make less and less CO_2 available for plants; thus, 95% of terrestrial plants will die progressively.

Only two types of plants will remain: first, those that carry out photosynthesis through the 4-carbon pathway (for example, corn) and not 3-carbon as is the case with the vast majority, and second, plants that carry out the acid metabolism of Crassulaceae such as cacti, pineapples, agaves and some algae *(Davis, 2019)*. These plants will continue to exist for another 400 million years.

In about 1 billion years, there will be no more plants on earth, and soon after, in the oceans, animal life will be the next kingdom to perish due to the cessation of oxygen production. The first animals to disappear will be large mammals, followed by birds and small mammals, amphibians, reptiles, and finally invertebrates. The little oxygen produced will come as it did at the beginning of life from bacteria. Specifically, cyanobacteria can photosynthesize even with 2-3 ppm of CO2 and tectonic activity. Other bacteria that will survive and probably the last in about 1.5 billion years will be anaerobes, especially those whose habitat is around submarine hydrothermal vents.

The appearance of us humans occurred when the timeline for life on Earth was already quite advanced. Approximately 60% to 70% of the time in which biology can exist has been consumed; if we consider that its beginning occurred some 3,800 million years ago. No more than 2,000 to 2,800 million years will probably be left when the last prokaryotes die.

In approximately 4.5 billion years, our star, the Sun, will deplete its hydrogen reserves faster and faster and reach a luminosity 40% greater than it currently has. The oceans will have evaporated several million years before, and even atmospheric water vapor over 150 ° C will escape into space.

Then, in 5.0 billion years, the Sun will begin to fuse helium to form carbon and oxygen. These reactions are more energetic than hydrogen, so the sun will transform into a red giant that would be bigger and hotter. By that time, it will already have engulfed Mercury and Venus. Finally, its perimeter will reach the Earth's current orbit, although, in those times, the Sun will have lost 30% of its mass. The decrease in its gravity will cause the Earth to move away from it several million kilometers, the temperature on our planet's surface will be so high that the rocks will melt, and the entire Earth's surface will be magma. Thus, the sun will continue to grow, and, in   ~7.59 billion years, it will devour the Earth *(Schröder, 2008)*.

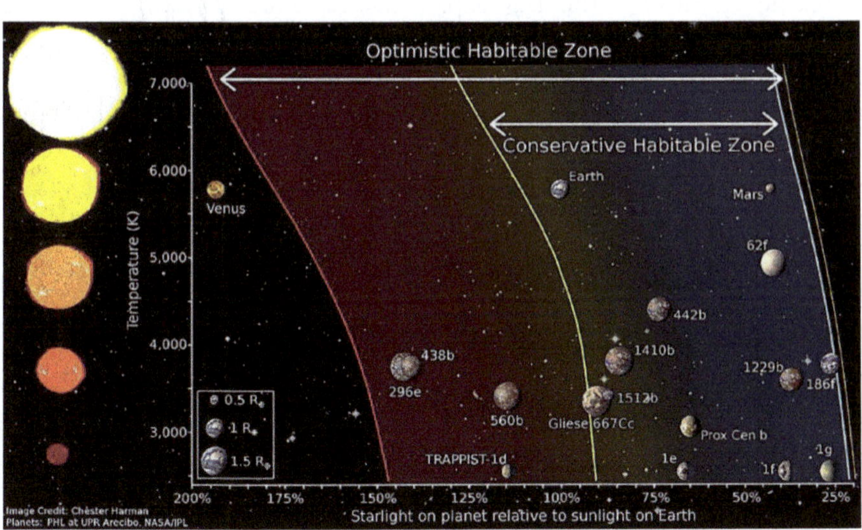

Figure 2.17   Circumstellar Habitable Zone.
Solar system planets and exoplanets.

Modified from: Chester Harman. CC BY-SA 4.0.

On the other hand, the Earth is very close to the inner edge of its circumstellar habitability disk (which we will talk about shortly). Our Sun, being a yellow dwarf star, is between 0.95 -1.37 Astronomical Units *(Fig. 2.17)*. Therefore, in 1.75 - 3.25 billion years, it will be so close to the sun that its distance will be less than 0.95 AU, causing the extinction of life before the sun becomes a red giant.

At the same time, the temperature on the moons of Jupiter, Ganymede, Europa, and Callisto, and on Enceladus -one of the moons of Saturn - *(Deamer, 2017)*, will gradually rise. Within the next 2.0 to 3.0 billion years, this elevation will be enough for them to have liquid water on their surface, likely allowing life on them to thrive.

For the rest of the Milky Way and other galaxies, the conditions necessary for life will decrease as entropy increases in the Universe. However, there will be enough energy to support life for at least 10 trillion (10^{12}) years if the end of the Cosmos occurs through the so-called Big Freeze (thermal death); or about 20 billion years if the outcome is through the Big Rip, *(Caldwell, 2003; Devlin, 2015; Loeb, 2016; Natarajan, 2019)*.

We will continue learning about other exciting concepts, including the Anthropic Principle, an essential part of this chapter. As an introduction, we will begin by asking ourselves a question:

Is life the purpose of the Universe?

Perhaps if we answered the question from a philosophical, religious, or even poetic point of view, the answer would undoubtedly be a resounding yes to the significance that this implies but let us see if scientifically based observations ratify religious and philosophical opinions.

The Universe is undoubtedly a site of transformation, where creation and destruction are intertwined. At the end of its life, the star that explodes as a supernova destroys its entire planetary system and even the surrounding stellar systems. Nevertheless, that enriched residual cosmic dust will contain new chemical elements. It will be the raw material for creating the next generation of stars with their worlds orbiting them and perhaps the appearance of new life.

If we observe the behavior of the Cosmos, we can realize that, in essence, it is neither hostile nor friendly, it is simply indifferent, and in our tiny cosmic outpost, the Earth, we observe it daily. For example, the cruel events of the food chain or the ravages produced by meteorological phenomena, which finally maintain life and renew the planet *(Davies, 2003)*.

The Universe probably has no purpose and only "exists or is," with laws in its origin that allowed the formation of chemical elements and, during its evolution, the construction of galaxies, stars, planets, and life. Life arises precisely thanks to the existing physical laws and the environment's characteristics surrounding it. This assertion is one of the significant concerns of astrobiologists who wonder if we will be able to recognize another form of life on a planet other than our own because we must admit that the only conception of life we have is the one shown by the provincial biology of planet Earth.

But let us continue with the initial concept —our Universe that gave rise to life will continue to evolve until its future physical characteristics prevent it from continuing to exist and disappear forever, giving rise to an ever colder and darker Cosmos. If we see it from this approach, then life turned out to be a mere circumstantial fact in the evolution of the Universe and not a purpose. At some point during the existence of the Cosmos, due to the physical constants with which it was born, it allowed giving rise to simple and even complex and intelligent life. Still, these same physical laws will prevent it from persisting in the Universe of the future. Life thus occupies only a part of the cosmic timeline.

Let us now look at what the Anthropic Principle developed more than 60 years ago tells us about our existence.

The Anthropic Principle

The participatory anthropic principle is an inverse thought regarding humanity's relationship with its Universe in which man does not exist as a function of the Cosmos, but rather our presence within it determines its nature and existence. This principle can be seen as a new way of placing the human being as the protagonist of the cosmic scene. The geocentric cosmology of Ptolemy represented for many centuries the belief of our preponderant role in Creation and satisfied the urgent need to transcend as the masterpiece of God; that is why it was so popular. Copernicus had to be very careful in exposing his new theory, and Galileo, supporting him, paid with his freedom and almost his life defending the heliocentric theory. However, in a way, they were both wrong; the sun is not the center of the Universe as established by William Herschel in the 18th century, not even of the Milky Way, as Harlow Shapley also demonstrated in 1918. Our galaxy is not just one among more than 2 trillion (2×10^{12}) that make up our observable Cosmos *(Conselice, 2016)*.

In a Universe with the characteristics we have described, our existence does not give us any prerogative to think that an anthropocentric philosophy can be correct and applicable. Nor would it be for any intelligent alien life form with characteristics similar to ours or even superiors. Even the image of the Creator fades as we learn more about the nature of the Cosmos, gradually becoming just a spiritual hope. So, centering the finality and magnificence of the Universe on humanity sounds like a joke in bad taste.

However, although this reasoning seems well-founded, we must consider an eclectic approach that, as a conciliatory principle of opposing philosophical opinions, it is possibly one of the best ways to approach the truth, contrary to the rigid ideological polarization. In which, we will give way to valuable opinions in favor of biocentrism and, as its derivative, anthropocentrism, whose core part is the Anthropic Principle supported by concepts of quantum physics.

"The Universe is like this because in it there are beings capable of wondering why it is like this." That is the essence of the anthropic principle; in other words, it is a philosophy that indicates that the observed characteristics of the Cosmos must be compatible with the conscious and sapience life that observes it.

This concept, proposed in 1973 by the Australian theoretical physicist Brandon Carter, is based on the theoretical concepts of Robert H. Dicke in 1961 and later supplemented through the *"Participatory Anthropic Cosmological Principle"* by John Wheeler, a physicist at Princeton and Johns Hopkins Universities, creator of the concepts of « black hole, wormhole, quantum foam, and white hole. » Wheeler suggested that *"observers create reality, and no phenomenon is a real phenomenon until it is observed."* These considerations derived from his delayed choice experiment *(Santaolalla, 2020)* and previous concepts such as Heisenberg's uncertainty principle and Schrödinger's cat paradox.

The uncertainty principle, developed in 1927 by the Nobel laureate in physics Werner Heisenberg, is a fundamental concept of quantum physics. It refers to the fact that it is impossible to simultaneously and precisely measure a particle's position and linear momentum (which is the product of mass times velocity). The more accurately we measure the speed of a particle, the more difficult it will be to determine its position in space and vice versa. The uncertainty principle is true only for the present and future times but not for the past, where we can measure both attributes accurately because they have already occurred. Translated into daily life, "our existence goes from a definite past to an uncertain future." Thus, quantum mechanics offers us a world in which exactitude and reality do not exist, only probabilities.

This statement gives rise to the paradox of the cat of the Austrian physicist Erwin Schrödinger who described it in 1935. The paradox is a mental experiment through which an attempt to transfer quantum mechanics to daily life by exposing the observer to superimposed situations, in this case, life and death, since the simultaneity of two quantum states is possible.

The paradox briefly described would be like this: a cat is enclosed in a steel box, and inside, there is a covered bottle filled with hydrogen cyanide under a hammer coupled to a device, which turns into a Geiger counter.

Figure 2.18 Schrödinger's cat experiment.

Modified from: fisicafundamental.net

Also, inside there is a container containing radioactive atoms with an exact probability of 50% that radioactive decay of an atom will occur in one hour. If this happens, the Geiger counter detects it and activates the device, the hammer will break the bottle, and the cat will inhale the poison and die *(Fig. 2.18)*. After one hour, the cat will have an equal chance of being alive or dead. Therefore, before the observer opens the box, the cat is in a state of life-death superposition.

In the Multiverse hypothesis, it would happen that, in some universes, the cat will be alive when the box is opened, but in others, it will be found dead. This experiment exemplifies what Wheeler wanted to tell us: "no phenomenon is a real phenomenon until it is observed," and consequently, "observers create reality.[26]

Now we return to the initial concept of the anthropic principle. While Brandon Carter was in Krakow, Poland, at the Symposium for the celebration of the 500th Anniversary of the birth of Nicolaus Copernicus, he presented a lecture, *"Great Numerical Coincidences and the Anthropic Principle in Cosmology" (Carter, 1974)*, where he pointed out that: « although our situation in the Cosmos is not necessarily central, it is inevitably privileged in a certain way. » The principle is controversial and has historically had supporters and opponents.

There are three main versions of the anthropic principle (weak, strong, and final). The first two were modified in 1986 from Carter's original concepts by physicists John D. Barrow and Frank J. Tipler in their book "The Anthropic Cosmological Principle" *(Barrow, 1986)*.

- Carter's weak anthropic principle indicates that: "our position in space and time is necessarily privileged since it is compatible with our existence as observers."

- Barrow and Tipler's weak anthropic principle argues: "There are places in the Universe where carbon-based life has developed because the Universe is old enough for this to have occurred."

---

[26] The theory of multiple or parallel universes, currently known as the Multiverse Theory, was first proposed in 1957 by the American theoretical physicist Hugh Everett III. Max Tegmark and Brian Greene devised classification schemes for the various theoretical types of multiverses in 2003.

- Carter's strong anthropic principle says: "the Universe and therefore the fundamental parameters on which it depends, must have the characteristics that allow the creation of observers within it at some point."

- The strong anthropic principle of Barrow and Tipler indicates that "the Universe must have properties that allow life to develop at some point in its history."

- The final anthropic principle of Barrow and Tipler refers: "a mode of intelligent information-processing must come into existence in the Universe (that is, intelligent life or artificial intelligence) and, once it appears, it will never disappear."

- Wheeler's Participatory Anthropic Principle says: "The existence of observers gives rise to the existence of the Universe."

These are the foundations of the Anthropic Principle with all its variants, and we can accept them or not.[27] The astrophysicist Lawrence Krauss thinks this principle is more philosophical than a scientific idea. However, we inhabit a Universe in which it has been developed life with intelligent observers capable of questioning their existence and the purpose of the Universe. However, we must be careful with how we reason our relationship with the Cosmos, its origin, and ours.

There is a great duality in philosophical-scientific thought. Whether our Universe is carefully planned from the beginning or whether, being one more of the infinite Universes that together make up the Multiverse, it was simply the fortuitous fact that its laws of origin and evolution allowed the appearance of life as we know it. However, there are so many coincidences or random events concatenated for the arrival

---

[27] The late British physicist Stephen Hawking suggested that our Universe is much less "special" than the Anthropic Principle proponents pretend. According to Hawking, the probability that a universe like ours arises from a Big Bang is 98% using the basic wave function of the Universe as the basis for his equations. Furthermore, Hawking has concluded that such a Universe can exist without any relationship with anything before it; that is, it could arise out of nowhere *(González dePosada, 2004)*.

of life on our planet (as we will see later). So, it is practically impossible not to be tempted to think that a higher order had regulated things *(Craig, 1988; González, 2004)*.

Scientific philosophy has played a decisive role in the evolution of humanity's thought, sometimes at the cost of the lives of its representatives, such as Giordano Bruno, but other times under the admiration and gratitude of its contemporaries and extemporaneous ones.

The conviction in the existence of a Creator and His designs have made more than one of the great geniuses hesitate from the results of scientific evidence. One of Einstein's most famous phrases uttered in his lectures in Brussels in 1927 is: « *God does not play dice,* » later written in a letter in 1944 addressed to Niels Böhr concerning the strict laws that govern physical phenomena of the Universe and to which the atoms are supposedly subjected by an already established order thus defending Laplace's deterministic theory[28]. However, quantum theory indicates the opposite, starting with the Heisenberg uncertainty principle, continuing with the butterfly effect as a typical example of quantum fields, and ending with several recent international experiments supporting and proving it.

In July 2004, during a conference, the late Dr. Stephen Hawking (1942-2018) stated, « *God not only plays dice but also sometimes throws them where they cannot be seen.* » In one of his last books, "The Great Design,"

---

[28] Niels Böhr (1885-1962), a Danish physicist, made fundamental contributions to the understanding of the atomic structure and quantum theory, for which he received the Nobel Prize in Physics in 1922. He had enormous discussions with Einstein, who was always a detractor of quantum mechanics and its orthodox or Copenhagen interpretation. In response to Einstein's letter, he wrote: Stop telling God what to do! Heisenberg recounts, recalling those discussions, that at one point, Paul Ehrenfest said: "Einstein, you embarrass me; you argue against quantum theory as your adversaries against relativity."

The basis of the determinism of Pierre-Simon, Marquis de Laplace of the 19th century lies in the phrase «if all the forces, positions and speeds of all the bodies in the universe were known, the future would be explained. »

Hawking and Leonard Mlodinow make a deep analysis and conclude by delineating a Universe in which a god is not necessary for its existence. Since time began at the moment of the Big Bang, nothing or anyone could have existed before creating it *(Hawking, 2010)*. However, this foundation is controversial with the theoretical physicists who support the Multiverse hypothesis and call it the "myth of the beginning of time." Hence, with this position, the debate about a Creator continues, and for religious conceptions, time is one more element of creation.

Humility is not one of the qualities of most scientific community members of the world. Still, some groups have suggested that perhaps we are not, or at least we are not yet smart enough to understand the intricate threads that constitute the fabric of the Cosmos. After all, 200,000 years of evolution of the terrestrial animal species with greater rational intelligence are not enough to ensure that we are ready to understand everything.

If we are objective, we must recognize that we know very little about our Universe, but we understand that it has specific characteristics for life to have developed. In the Multiverse, there is enough space to imagine all kinds of Universes. From those that only last a fraction of a second to those that will continue their existence forever. Those in which cosmic structures different than stars and galaxies are formed and those in which there are only a few subatomic particles analogous to ours. Those in which life is superabundant and others in which there will never be living beings. Finally, universes with different and unimaginable physical laws for us, with several physical and temporal dimensions greater than ours. Nevertheless, we are lucky to exist thanks to the characteristics of the Universe of which we were part.

Nevertheless, if the physical laws governing us were slightly different, for example, if the strong nuclear force were only 20% weaker relative to the electromagnetic force, there would be no chemistry, no biology, and only hydrogen. Alternatively, if only 2% stronger, no hydrogen would exist; only heavier elements would exist, and water could not be formed. If the force of gravity were stronger, the existence of the stars would be shorter, and there would only be time to develop simple forms of life and probably non-complex.

If the cosmological constant had been a little larger at the birth of the Universe, then stars and galaxies could not have formed since the accelerated expansion would not have allowed it *(Hawking, 1988; Davies, 2003).*

Finally, we know that the mass of the neutron is slightly greater than that of the proton (0.14%); this leads to the rapid decay of the (free) neutron into a proton, which is more stable. If, on the contrary, the proton decayed into a neutron and the fusion reactions were based on neutrons, naked nuclei of all masses would be formed.

The only material in the universe would then be neutronium, chemical elements could not constitute, and life could not develop *(Wilczek, 2015).* As we view our Universe, it is delicately tuned for the emergence and sustenance of life, at least for a time.

Let us now see more about the essential requirements for life to emerge on our planet.

Fundamental Components for Planetary Life

In 1584 the astronomer and philosopher Giordano Bruno (1548–1600), considered the Father of Astrobiology, published his book *"On the Infinite Universe and its Worlds." —*In it, he writes: *"There are innumerable suns and countless planets earth that revolve around these suns in the same way that the planets revolve around our Sun. Living beings inhabit these worlds"* (Fig. 2.19).

Figure 2.19   Engraving by Camille Flammarion in her work:
"L'Atmosphere: Météorologie Populaire," from 1888.
At the bottom of the illustration reads: "A medieval missionary says
that he had found the place where Heaven and Earth touch."

Source: Public Domain.

Likewise, in the scholastic texts of Abhidharma (3rd century BC), where the sutras or teachings of Buddha (Shakyamuni), Siddhartha Gautama, are recorded, we can read in one of them: *"The Universe contains innumerable systems of worlds, and in many of those systems a Buddha is performing beneficial actions for spiritual practice"* (Landaw, 2011).

Various attempts have been made to mathematically elucidate the possibilities of life in our Galaxy for more than sixty years. In 1961 the radio astronomer Frank Drake developed an equation that bears his name to calculate the number of planets with intelligent beings in our galaxy and that, for example, would be able to broadcast radio emissions detectable in ours, probably with the wavelength of hydrogen which is 21 cms equivalent to 1420.4 Megahertz, since being the most common element in the Universe, it would be feasible that alien intelligence would consider it logical as a form of transmission within the radio spectrum *(Drake, 1961)*.

Frank Drake Equation
$$N = R \cdot f_p \cdot n_e \cdot f_l \cdot f_i \cdot f_c \cdot L$$
$N =$ the number of technologically advanced.
$R =$ the total number of stars in the Milky Way.
$f_p =$ the fraction of those stars that have planetary.
$n_e =$ the number of planets suitable for life, in each planetary system.
$f_l =$ the fraction of those planets where life develops.
$f_i =$ the fraction of those planets where intelligence.
$f_c =$ the fraction of those planets with communication by radio signals.
$L =$ the fraction of the planet's lifetime during which civilization lives.

Table 2.1 Frank Drake Equation. SETI Institute.

This equation *(Table 2.1)* identifies some of the specific factors assumed to have an essential role in developing civilizations *(Vakoch, 2015)*. Historically, estimates have been made of the possibility of planets with "intelligent life that can communicate via radio signals in the Milky Way," ranging from 1 to 50 million. Drake's original calculation resulted in 10 planets; Various calculations based on recent data place the number highly likely to be less than the original. However, most agree it is very close to one *(Forgan, 2009; Seager, 2018)*.

In 2000 the book *"Rare Earth: Why Complex Life is rare in the Universe"* was published. Peter Ward, a geologist, and paleontologist, and Donald E. Brownlee, an astronomer and astrobiologist, describe several previously not included factors. For example, the Rare Earth hypothesis is contrary to the "principle of mediocrity" proposed in the equation of Franz Drake and the Order of the Dolphin, of which Carl Sagan was a part.

From their point of view, planets, planetary systems, and galactic regions that are friendly to complex life, such as Earth, the Solar System, and our area in the Milky Way, are probably exceptional. In this way, they conclude that simple microscopic life can be relatively abundant in the Universe. However, complex life is rare, and the Rare Earth hypothesis is the probable solution to the Fermi Paradox that says: *'If extraterrestrial civilizations are common. Where are they, and why aren't they obvious?*

Based on this hypothesis, complex life's appearance requires exceptional circumstances not stated in the Drake equation. For example, a rocky planet must have a specific size rank, possess magnetosphere and tectonic plates, and specific chemical characteristics of its atmosphere. In addition, its lithosphere and oceans must have boosters of evolution, such as glaciations and impacts of life-transporting meteorites, and particular factors that caused the explosion of the animal *fila* in the Cambrian ~540 million years ago *(Kipp, 2017)*.

However, the emergence of intelligent life may also require exceptional and random events. Many of the necessary variables fall within very narrow ranges. However, the vastness of the Milky Way and even more so of the Universe is such that it contains many planets similar to Earth, separated by many light-years apart, preventing communication among their civilizations. So that explains the *Fermi Paradox*.

<table>
<tr><td colspan="2" align="center">**Rare Earth Equation**</td></tr>
<tr><td colspan="2">$$N = N^* \cdot n_e \cdot f_g \cdot f_p \cdot f_{pm} \cdot f_i \cdot f_c \cdot f_l \cdot f_m \cdot f_j \cdot f_{me}$$</td></tr>
<tr><td>N^*</td><td>the number of stars in the Milky Way.</td></tr>
<tr><td>n_e</td><td>the average number of planets in the "habitable" zone of a star.</td></tr>
<tr><td>f_g</td><td>the fraction of stars in the habitable zone of the galaxy.</td></tr>
<tr><td>f_p</td><td>the fraction of stars with planets in the Milky Way.</td></tr>
<tr><td>f_{pm}</td><td>the fraction of rocky "metallic" planets.</td></tr>
<tr><td>f_i</td><td>the fraction of habitable planets where microbial life can arise.</td></tr>
<tr><td>f_c</td><td>the fraction of habitable planets where complex life evolves.</td></tr>
<tr><td>f_l</td><td>the fraction of the total planetary existence during which complex life is present.</td></tr>
<tr><td>f_m</td><td>the fraction of habitable planets with a large moon.</td></tr>
<tr><td>f_j</td><td>the fraction of planetary systems with Jupiter-like planets.</td></tr>
<tr><td>f_{me}</td><td>the fraction of planets with few extinction events.</td></tr>
</table>

Table 2.2 Rare Earth Equation. Peter Ward y Donald Brownlee.

In this equation, all the elements come from components that exist in our galaxy, solar system, and especially on our planet. Learning from them allows us to become aware of the enormous number of coincidences and sequenced events that made us have the fortune to exist, to be alive here and now.

Next, we will briefly review each of the Equation components that constitute the *Rare Earth Hypothesis (Table 2.2)*.

(N *) The exact number of stars in the Milky Way is difficult to know. It estimates a range from 100 billion to one trillion, with a higher probability of being between 200 to 400 billion. According to the High Accuracy Radial Velocity Planet Searcher and the European Southern Observatory, around 75% of the stars in our galaxy are red dwarfs. Most of these stars have rocky planets orbiting them, so there must be tens of trillions of these planets in our galaxy. However, although 40% of these planets are within the circumstellar habitable area, red dwarf stars are characterized by frequent stellar flares or eruptions, which bathe their planets with high doses of X-rays and ultraviolet radiation. It makes the origin and evolution of life difficult, at least on its surfaces. In contrast, only 10% of the stars in the Milky Way are yellow dwarfs like our sun, which are probably the most appropriate to allow life.

(ne) Planetary habitability is a measure of the potential of a cosmic body to support life and can be applied to both the planets and the natural satellites of the planets. The circumstellar habitability zone is called the annular region around a star in which the luminosity and the incident radiation flux allow the presence of water in a liquid state on the surface of any rocky planet or satellite that has a mass between 0.5 and 10 earth masses and an atmospheric pressure greater than 6.1 mbar *(Fig. 2.20)*. This area is called the *"Goldilocks zone,"* whose principle is, as in the story, that extremes should be avoided (not too hot, not too cold, not too big, or too small). The planets' distance that orbits their stars must be in the habitable zone and varies according to the stellar spectral class.[29]

---

[29] Based on its surface temperature, the stellar spectral classification currently used with some modifications is the Morgan – Keenan. They are named starting from the warmest to the coldest with the capital letters O, B, A, F, G, K, M, L, and T. Each of the classes also has a sub-classification with a number ranging from 0 to 9, also from the hottest to the coldest. For example, our Sun is classified as a G2 star. The stars classes L and T are brown stars, typically cold and with very little luminosity.

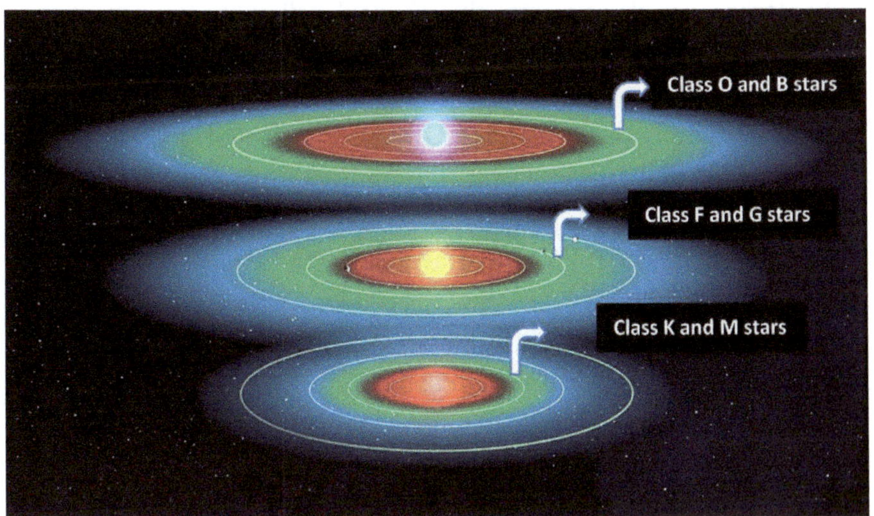

Figure 2.20   Circumstellar habitability zones (in green).
Modified from NASA public domain. CC BY-SA 3.0.

As we have already pointed out, for red dwarfs, 40% of their rocky planets are in that area relatively close to their star, especially if it is tiny. The best representative is the *Trappist-1 system* which is 39 light-years from us in the Aquarius constellation, with seven rocky planets orbiting its red dwarf star, which is the size of Jupiter but 85 times more massive. Three planets (e, f, g) are found in this planetary habitable zone. However, there may be no life in any of them because the magnetic storms of their star that are between 100 and 10,000 times more powerful than our sun have indeed destroyed their atmospheres *(Vida, 2017)*.

The same has happened with the planet *Proxima b*, the rocky exoplanet closest to Earth, 4.22 light-years away. Its star Proxima Centauri, a red dwarf (class M), produces impressive energy, ultraviolet light, and X-rays that ionize the atmosphere creating a strong escape effect of oxygen, hydrogen, and nitrogen into space, which prevents liquid water from forming on its surface and an atmospheric pressure compatible with life *(Airapetian, 2017)*.

In larger stars like our sun, the *"G-class"* yellow dwarf stars, one in five (18%) have rocky planets in their habitable zone *(Kunimoto, 2020)* with a size and temperatures similar to Earth, the closest -*Tau Ceti*- just 12 light-years away from us. These are the star systems where complex and intelligent life is most likely to be found, and it is estimated that about 6 billion of these planets are in the Milky Way. On the other hand, the formation of life on planets around massive blue and white stars in their habitability zone (spectral classes O, B, and A) is less likely since they tend to have more unstable and violent behaviors. Furthermore, although their habitability zone is much more distant than red and yellow dwarfs, the ionizing radiation can prevent the beginning and evolution of life unless these planets have an extremely intense magnetosphere. Still, on these planets, the photo-evaporation phenomenon occurs, which is the process by which, due to the intense solar wind, they are stripped of their atmosphere.

Furthermore, the lifespan of these stars is relatively short. Moreover, the diversity of the chemical elements they contain may be poorer, so even though they could have been achieved, they would only have reached the time to develop simple life. In such a way, the efforts to find complex and intelligent life on the planets are focused on less massive and hot stars, especially those of class F, G, and K (white-yellow, yellow, and orange stars). The term "yellow dwarf" is a misnomer for G-type stars. In fact, the Sun is white but can often appear yellow, orange, or red from Earth due to the phenomenon of atmospheric Rayleigh scattering.

(fg) The galactic habitable zone forms a ring around the galaxy's center *(Fig. 2.21)* which ranges from the center of the Milky Way, from 4 kiloparsecs (13,000 light-years) to 10 kiloparsecs (32,600 light-years). It comprises the only region in which stellar systems have planets where life can appear. This zone is located between 7 and 9 kiloparsecs for planets similar to Earth. However, beyond its outer limit, the metallicity of the stars is too low to allow the formation of rocky planets like Earth, and closer to the violent galactic center, exposure to highly energetic events like gamma-ray bursts makes it very hostile to life[30] *(Gowanlock, 2011).*

*Figure 2.21   Milky Way - galactic habitable zone (green)
and position of the sun.*

Modified from: NASA public domain. CC BY-SA 3.0.

(fp) The fraction of stars with planets in the Milky Way could theoretically be close to 100%. The birth of a star implies being in the center of a gas and dust *disk* called *protoplanetary*; much of the gas and dust is swept away by stellar radiation. However, rocky and gaseous planets can form around their central star before this happens. A new

---

[30] In astronomy, the term "metallicity" refers to the presence of elements heavier than hydrogen and helium, such as oxygen, silicon, carbon, iron, and others.

analysis of the Kepler data shows that one in six stars has a planet the size of Earth, equivalent to 17 billion. About a quarter of all the stars in the Milky Way have a super-Earth, and in the same fraction, there is a mini-Neptune. Only about 5 percent of the stars have a large Neptune, and a little over 6 percent a gas giant, most at very close orbital distances from their star and are called "hot Jupiters" *(Fig. 2.22)*.

Figure 2.22 Percentage distribution of exoplanets
Credit: Francois Fressin. Harvard-Smithsonian Center for Astrophysics.

Meibom, S: (2013). The same frequency of planets inside and outside open clusters of stars. *Nature*, 499.55-58.

(fpm) Approximately 40% of the planets found in the habitable galactic zone correspond to the fraction of rocky "metallic" planets. The galactic habitable zone corresponds to one-third of the galactic radius, covering an area close to 20,000 light-years; therefore, the minimum number of rocky planets in this area of the Milky Way is 30 billion. Of these, very few could harbor life at some point in their evolution. Most of them, among other peculiarities, conserve their original dense atmosphere formed by hydrogen and helium, whose characteristics prevent biology from appearing and suffocating any life settled on their surface. These planets have been called *«inhabited habitable worlds. »* Locating planets with life outside our solar system is currently impossible since detecting their bio-signatures (which refer to any element, isotope, molecule, or phenomenon that provides scientific evidence of the past or present of life) is still out of reach *(Cockell, 2014).*

(fi) It is the fraction of habitable planets where microbial life can arise. This component of the equation and the following components are purely speculative since we do not have any accurate information. However, we can imagine with a good chance that a low percentage of those 30 to 40 billion planets have the physical and chemical conditions necessary for forming protocells and other simple single-celled life forms.

Astrobiology is an interdisciplinary science derived from the biological sciences and the Earth and space sciences. Russian astronomer Gavriil Tikhov coined the term in 1953, and its goal is the search for life on other planets or satellites.

Life on all planets is defined based on specific theoretical bases, such as the necessary elements of the Periodic Table that a complex system must contain to consider it "alive." Six essential elements are grouped mnemonically with the terms « SPONCH or CHNOPS » C – carbon, H – hydrogen, N – nitrogen, O – oxygen, P – phosphorus, and S – sulfur. Thus, the presence of these six elements on any planet's surface is a solid potential indicator of life.

However, precisely with the planet Mars for many years, there was a doubt whether it could have had life or not because none of the robotic missions since 1976 (Viking 1 and 2) had been able to identify the existence of nitrogen. Not until April of 2015 did the Curiosity mission identified traces of nitric oxide (NO) and hydrocyanic acid (HCN).

Another of the fundamental elements of Astrobiology to identify life is the presence of certain gases in the atmosphere, such as carbon dioxide, methane, oxygen, and ozone. These are final or intermediate products of metabolism, photosynthesis, or fermentation. Each element is a biosignature and spectroscopy of how these life traces can be found. Modern Astrobiology was born in 1959 as NASA's first project and program. In the past, it received the names —Exobiology and Xenobiology; however, Astrobiology is the most used today.

(fc) It is the fraction of planets where complex life evolves. This component of the equation is even more difficult to calculate. Our world lacked multicellular life for its initial 2.7 billion years of existence. The first multicellular eukaryote was *Grypania spiralis,* a spiral ribbon-shaped fossil 0.2 to 2 millimeters wide and 1 to 10 centimeters long that appeared about 1.87 billion years ago. The first animals, *ctenophores,* appeared just 610 million years ago as part of the so-called Ediacaran organisms with soft bodies and lacked skeletons, similar to sea nuts and current- jellyfish *(Baxevanis, 2013)*. However, roughly 540 million years ago, sizable diversity of animal life finally emerged in the oceans. It did so relatively suddenly and unexpectedly in an event known as the Cambrian explosion. Until a few years ago, the cause of this extraordinary phenomenon was unknown.

In November 2017, Michael Kipp and Eva Stüeken from the University of Washington and the NASA Astrobiology Institute published in the *Journal of Science Advances* on what could explain this explosion in the diversity of the Cambrian animal *phyla (Kipp, 2017)*. They discovered that the amount of phosphorus available in the oceans was very low until 800 million years ago, coinciding with the increased ambient oxygen concentration necessary to recycle this element properly.

Phosphorus is one of the six essential elements for life to develop (RNA, DNA, phospholipids). In addition, it is the only element capable of storing and transferring energy and forming molecules such as ATP (adenosine triphosphate).

The low amount of planetary phosphorus available for hundreds of millions of years led to low biological productivity.

Figure 2.23 *Artistic reconstruction of the «Tiktaalik, » the first land animal, its predecessor Panderichthys and its successor Acanthostega.*

Modified from pengembarabijaksana.blogspot.com

The phosphorus concentration in the Precambrian period was ten times lower than today. Therefore, as its concentration gradually increased, it achieved the necessary recycling of the element from plankton (which were the first life forms to capture it) towards the new animal forms that now did have the possibility of becoming more complex.

This phosphorylation that occurred like the "great oxidation" caused about 540 million years ago, a determining effect on animal variability in the oceans, and later on, land 375 million years ago, when the first land animal appeared, a tetrapod fish with lungs and gills called

Tiktaalik roseae, 2.70 meters long with an intermediate constitution between fish and crocodile *(Fig. 2.23)*. Subsequently, this animal genus emerged from amphibians, reptiles, mammals, and birds.

From a planetary point of view, researchers from the University of Cardiff recently suggested that there may be a cosmic deficiency of the essential phosphorus for life. Phosphorus is formed from the supernovae explosion and then travels through space in meteorites until it hits the planets, but apparently, not all supernovae produce the same amount. Some make very little, and the concentration of this element is different in various regions of the galaxy. So that some planets may lack or have a low concentration of reactive phosphorus due to their area of origin. In that case, life could have great difficulties starting and especially evolving in those worlds[31]. *(Greaves, 2018).*

(fl) It is the fraction of the total planetary life during which complex life is present. Finally, this is the component of the most complicated equation to calculate. Our planet lacks evidence of complex life for 88% of its existence, so probably unicellular and simple multicellular life are the best representatives of life on Earth, especially the former. The human being occupies less than 0.005% of the time in the history of the Earth. In such a way that, to project this component in our galaxy, it would probably have to be similar to our chronology, considering individual variations. Once it exists, the permanence of intelligent life will depend on two main factors; the first is how long it takes since it appears until it self-destructs. The second would be the last component of this equation: the presence of local natural events or galactic capable of causing mass extinction, which tend to be random for each planet.

---

[31] During the formation of the Earth, meteorites provided an important source of phosphorus. However, it was later the billions of rays over hundreds of millions of years that, when falling to the ground on volcanic rocks, transformed the insoluble phosphorus into hydrosoluble making it available for the start of life and later making its complexity possible when ambient oxygen was increased *(Hess, 2021).*

(fm) The answer to the fraction of habitable planets with a large moon we do not know. There are moons even larger than ours in the solar system, such as Ganymede and Callisto, that orbit Jupiter, and Titan revolves around Saturn. However, these are gas giant planets and therefore are considered non-habitable (although we do not have absolute certainty), but no other known rocky planet has them. The current technological capacity has made it possible to observe some extrasolar planets similar to Earth, but it cannot identify natural satellites that orbit them. However, since the Kepler Space Telescope Mission Task Force calculation in our galaxy reports billions of Earth-like planets, it is pretty sure they exist.

Theia's impact on the Earth at the time and how it occurred was a rare phenomenon. For example, if the impact had been on the opposite side, the Earth's rotation would have been retrograded, as it must have happened to Venus, resulting in a more extended rotation period than the orbital one. Thus, we would not have a moon either. If the enormous impact had occurred at a later stage in the formation of the Earth, the greater mass and gravity of the planet would not have allowed enough matter to be expelled to form a large moon. If the impact had occurred millions of years earlier, due to the lower mass of the Earth, much of the debris would have been lost to space, and the resulting Moon would have been too small to stabilize the obliquity of the rotation axis of the Earth. If this mega-impact had not occurred, the Earth would have conserved a more significant amount of water, carbon, nitrogen, hydrogen, and helium. Perhaps this would have led to a greenhouse effect that would make it impossible to develop complex life.

Of the many elements of the Rare Earth hypothesis, the presence of our giant moon appears to be one of the most important and puzzling. Without the great Moon, the Earth would have had a volatile atmosphere, and it seems unlikely that life would have progressed as successfully as it did.

Even with Earth's relatively stable long-term climate, it took 92% of its existence for land animals to develop. Thus, the Moon, our closest neighbor in space, has played a prominent role in the origin and evolution of life on Earth. Its presence is particular and unique *(Ward & Brownlee, 2000).*

(fj) According to current results, of the more than 830 multiple planetary systems found to date, approximately 6% have Jovian planets, also called gas giants *(Multiple Planets systems, 2022).*

The existence of Jupiter from its time and place of formation has had a profound influence on the Earth's ability to generate and maintain a stable environment for life. Jupiter is nearly two and a half times more massive than all the planets in the solar system, and it would fit more than 1,300 Earths.

Due to its enormous gravity that cleanses our solar system from dangerous asteroids and comets that could cross the Earth's orbit, Jupiter has had an extraordinarily beneficial influence on life on our planet, as we were able to verify in 1994 when Comet Shoemaker-Levy 9 collided with it. The comet was fragmented into 23 pieces millions of miles away before impact due to the gas giant's enormous force of gravity. Several fragments left marks on the planet's surface with a diameter greater than Earth's. If any pieces had struck Earth, there would have been a Permian-Triassic mass extinction. Jupiter is known to be the oldest planet in the solar system. Specific studies with tungsten and molybdenum isotopes concluded that its formation was less than 10 million years after the sun *(Kruijer, 2017).* Due to its rapid formation, Jupiter acted as an effective barrier against the internal transport of planetary material through the disk, which could partly explain why our Solar System lacks super-earths[32].

There are many reasons to suspect that our solar system originally housed more and larger planets, as in our galaxy's hundreds of newly discovered multi-planetary systems. Although there is nothing in our solar system between the Sun and Mercury, these equivalent regions around most of the other stars appear to be filled with planets of

---

[32] A super-earth is a rocky planet with more than one but less than ten landmasses.

medium mass and sizes between Earth and Neptune called "super-Earths" with periods of very short orbitals. Now that we can compare our planetary configuration to the structure of most other systems, our solar system appears a "weirdo." There is no reason to suspect that the planet formation process around our sun was different from that of other stars. Instead, the explanation for the atypical state of our solar system could lie in the details of its subsequent evolution, mainly controlled by Jupiter. Let us analyze this.

For decades, the planetary systems were considered reasonably stable; rocky planets would form in the intense light and heat near stars, while gas giants would originate farther away, where cold temperatures better preserve gases. All they would revolve around their star in nearly circular orbits. However, current information seems to contradict this assumption. Some planets in other systems do not orbit their star in tight circles but instead follow wide elliptical orbits due to gravitational influence with other worlds or to orbiting binary star systems. Many gas giants discovered planets orbit much closer to their sun than Jupiter and Saturn. Furthermore, many of these extrasolar gas giants might have migrated shortly after their formation, driven by interactions between their proto-planetary disks and their star or by proximity to their neighboring planets *(Billings, 2015)*.

Based on these observations, some groups of astrophysicists are investigating planetary migration and have designed some hypotheses such as the *"Grand Tack."* The theory postulates that, in the first million years of our solar system, Jupiter, shortly after its formation at 3.5 astronomical units (AU), migrated inward up to 1.5 AU and then returned to the outside of the solar system. At that time, Jupiter still had a gaseous disk. Due to gravity's force, much of that gas was drawn in a spiral toward the Sun and Jupiter, entering the orbit that currently occupies Mars.

Jupiter would have continued to fall towards the Sun if it had not been for Saturn, which also began to drift. As the two giant planets got closer, they were caught in a 2:1 orbital resonance that ejected the gas between them, gradually reversing their spirals and causing them to veer back toward the outer solar system. The physical mechanisms underlying the "Grand Tack" hypothesis are robust, and there is good reason to believe it did. The theory also clarifies the abnormal small size of Mars, which should be more prominent due to a large number of planet-forming materials in its orbital region at that time. According to this hypothesis, Jupiter would have attracted most of that material, leaving what Mars needs to form with its dimensions. The view also helps to explain the distribution of the rocky bodies that make up the Asteroid belt and the hypothetical population of super-proto-Earths that were pulled inward by the Sun's gravity due to the influence of Jupiter.

After Jupiter's migration, only remnants of volatile gases and shattered rock residues remained, which Jupiter propelled into the solar system with enough material to have formed Mercury, Venus, and Earth in 100 to 200 million years. Instead, Mars probably did it in less than 10 million years. That is consistent with much evidence suggesting that the rocky planets in our inner solar system formed later than the outer gas giants and explained why the planets closest to the Sun are smaller and have thinner atmospheres than their counterparts around other stars *(Batygin, 2015)*.

Our solar system does not look like the rest of the discovered star systems. It indicates that we could still be more alone than previously thought, and truly Earth-like planets, with solid surfaces and moderate atmospheric pressures, are rare. If this is confirmed, it would mean that the vast majority of nearby, potentially rocky, habitable planets that we now observe around so many stars might not be rocky or habitable. In such a case, on those worlds, any form of life would be crushed, burned, and suffocated under their thick hydrogen-saturated atmosphere.

The discoveries also suggest that it is infrequent to find planets such as Jupiter so far from their star, which continue to migrate inward instead of returning to outer orbits, interfering with the formation of

Earth-like worlds. The bottom line from this Grand Tack hypothesis is that it is Saturn that we have to thank for being here. The existence and location of Saturn prevented Jupiter from staying closer to the Sun; therefore, the Earth could be formed as we know it.

(fme) The fraction of planets with few events that cause extinction is unpredictable. The rule is that all the worlds are exposed to violent and powerful forces during their formation. Some rocky planets become incompatible with life due to extreme and permanent changes in temperatures and pressures or titanic collisions of meteorites and comets such as those that occurred between 470 to 770 million years after the formation of the Earth during the *"late heavy bombardment period."* They could leave craters up to 5,000 kilometers in diameter, now evident on Mercury and our Moon *(Gomes, 2005)*. In the last 541 million years since the Cambrian period, more than 20 mass extinction events have occurred, five of which have been particularly catastrophic. The causes include a hypernova explosion close to Earth (~6,000 light-years), intense and prolonged volcanism, and the collision of large comets or meteorites like the one that occurred in the last great extinction 66 million years ago, which killed the dinosaurs.

To conclude, we can affirm that the *"Rare Earth"* equation allows us a much more solid basis for the probabilities that simple and complex life exists on certain planets of the Milky Way. Compared to *"Frank Drake,"* the equation necessarily forces us to reflect on the fortune we have to live and the responsibility we acquire towards our planet and all its forms of life by being the dominant species.

Dr. Jack W. Szostak, Nobel Prize in Medicine, observes: *"Is our Earth the only planet that hosts life? The answer to this question depends on how easy or difficult it is for life to emerge from the chemistry of a young planet. If planets with the right environmental conditions are common, and the life path is a series of simple or high-probability steps, life could be abundant in the Cosmos. On the other hand, if there are several low probability difficult events on the life path or even a single extremely difficult step, then life could be very rare in the universe. In this extreme, our planet could be the only place in the Universe with life. Right now, we do not know. However, this is where astronomical studies intersect with laboratory studies. If evidence of life is found on a distant planet, then the fact that life arose twice independently would mean that there is no complicated step in the origin of life, and it could be easy for us to reconstruct a pathway to life in the laboratory. Conversely, if our work in the laboratory suggests that all steps from geochemistry to life are easy, the implication would be that life must exist on other planets and that we can be trusted to advance in its search elsewhere in the Universe"* (Szostak, 2016).

Conclusions:

1. From the cosmological point of view, our Universe is a place where life is possible without a guaranteed period. It arises and ends between cycles of creation and destruction, adapting to specific environments, many of them genuinely extreme. Even when they cause extinctions, catastrophic events are opportunities for the renewal and diversification of all vital domains. In this context, life is the great guest to the realm of existence under the rules of quantum theory, where there are only probabilities.

2. Essentially, biocentrism indicates that our Universe is finely tuned for the existence of life *(Dicke, 1961)*. The anthropic principle further adds that it is carefully designed for the existence of conscious intelligence. However, it must be considered that this type of intelligent life is possibly a mere accident of an evolutionary experiment. Hundreds of animal genera and species survived on Earth for tens of millions of years. However, they never managed to significantly increase their intellectual capacity, unlike a group of hominids that, in just 200,000 years, went from being nomadic hunters and gatherers to making space travel. Thus, history teaches us that nature has been much more interested in other means of survival beyond intelligence.

3. From the first *Homo sapiens* to the present, understanding the essence of the Universe and its treasure «life» has been the most significant achievement that humanity has had. The discovery of this gem in other

worlds is undoubtedly a fundamental goal to answering our most profound existentialist dilemmas, and also, why not say so, to feel a little less alone in the vastness of the Cosmos. In his book *"Contact,"* Carl Sagan quotes one of his best-known phrases: « *If We are alone in the Universe, then it is an Awful Waste of Space.* »

Origin and Evolution of Humanity

When in Chicxulub (which in the Mayan language means *devil's flea*), 66 million years ago, a meteorite ~12 kilometers in diameter (7.5 mi) fell in the Yucatan Peninsula, the dinosaurs were the dominant vertebrates. Since the Upper Triassic period, they had already reigned for more than 165 million years *(Nesbitt & Renne, 2013)*.

The local geography was not the same as what we witness today. The continental drift had not yet made up much of the territory of Mexico. Instead, the northern region of the Yucatan Peninsula was under the sea, as was the peninsula of Florida and Central America, as we can see in Figure 2.24 based on the Paleomap developed by Dr. Ron Blakey of the University of Northern Arizona *(Keller, 2011)*. This meteorite, 35% larger than Mount Everest, impacted the Earth in shallow waters southwesterly, with a 45 ° - 60 ° inclination at a speed of ~ 75,000 km/hour (47,000 mi/hr) *(Collins, 2020)*. With force equivalent to the explosion of 100 billion tons of TNT (1×10^{14} = 100 teratons), leaving a crater of 180 kilometers in diameter (112 mi) and 40 in-depth (25 mi) with the consequent formation of megatsunamis up to 280 meters (918 ft) high in the areas surrounding the impact, covering the coasts more than 300 km (186 mi) inland including some places that today make up the southeastern United States. In fact, there is evidence of these tsunamis up north until the coasts of New Jersey.

This impact is the cause of the Cretaceous-Paleogene mass extinction. As if that were not enough, at the same time, a colossal volcanic activity occurred on the Deccan Traps in west-central India, which lasted about a million years, with lava flows exceeding 1.5

million. km2 (932 mill sq mi) of extension and up to a thousand meters high. The extinction of the dinosaurs was sealed.

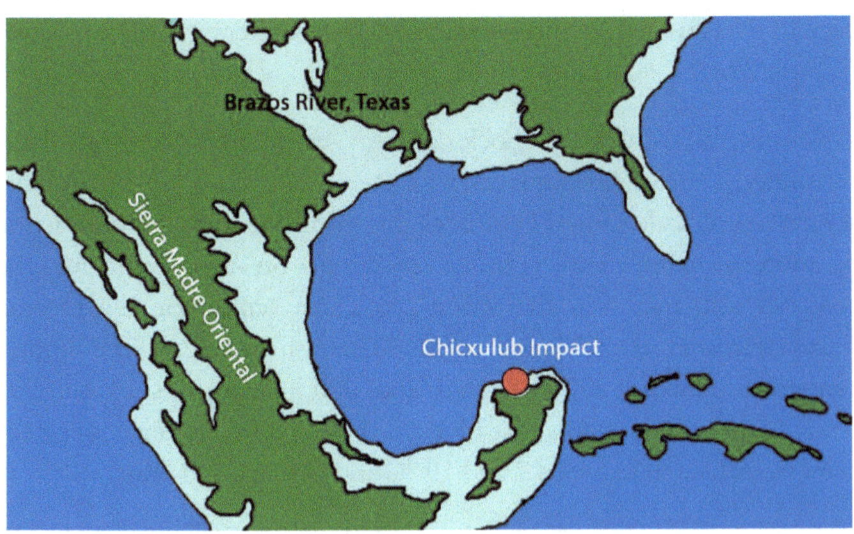

*Figure 2.24 Paleogeographic reconstruction of the
impact site at Chicxulub.*

Credit: R. Alejandro Betancourt
(Based on the Paleomap by Ron Blakey, R., NAU Geology).

Nevertheless, let us see how this mass extinction occurred. The impact zone was located in an area very close to the coast, the maximum sea depth was probably less than 200 meters, and the soil was made up mainly of gypsum rich in sulfur.

A gigantic cloud consisting of water vapor, gasified rock, and superheated dust to 10,000 °C spread at enormous speeds from the impact crater and rose beyond the stratosphere.

The expansive wave reached 900 km (560 mi) in diameter almost instantaneously and expelled the materials excavated together with incandescent pieces of the meteorite into the atmosphere. A percentage exceeded the escape velocity. It went out into space, but those that remained, called spherules, re-entered, burning vast areas of the continent's surface through multiple forest fires. These spherules fell in Patagonia only 10 to 15 minutes after the impact. However, the enormous and growing explosion cloud heating to thousands of degrees Celsius led to fires even 2,000 kilometers (1,250 mi) away *(Roberts, 2017)*.

Meanwhile, colossal shock waves were produced in large surrounding areas. Earthquakes well above 10 points on the Richter seismological scale caused landslides that contributed to the production of more tsunamis and volcanic eruptions on the continent; and probably also increased the gigantic volcanic activity on the Deccan Traps *(Richards, 2015; Byrnes, 2018)*. The ash raised by the impact contained between 200 thousand and 400 billion tons of sulfides, among other components that blocked the sunlight, rapidly cooling the planet's surface to between 7 and 10 °C and causing acid rain.

In weeks, this event broke the food chain of the oceans and land, both due to the change in the soil and water chemistry and the sudden interruption of photosynthesis caused by the death of 60% of the plants, conditioning the disappearance of 75% of the species of the animal kingdom; the predators probably being the last to die. None of the giant animals survived the extinction, and only the smaller ones did, including omnivores, insectivores, and scavengers, perhaps due to the greater availability of their food sources *(Twitchett, 2007)*. No exclusive herbivorous or carnivorous mammals survived. Instead, the surviving mammals, crocodiles, turtles, birds, and amphibians, feed on insects, worms, and snails, which in turn feed on dead plants, animals, and organic detritus *(Sheehan, 1986)*.

The massive death of the dinosaurs occurred very quickly. In the American Continent, the highest percentage of fatalities occurred on the same day of impact, and in the rest of the world, from weeks to months.

As the sky cleared after several years, the Earth suffered global warming caused by the greenhouse effect due to excess carbon dioxide and methane in the atmosphere. It probably contributed to the extinction of some species that had survived the tremendous impact and the prolonged winter produced by the cloud of gases *(Schulte, 2010)*. It is calculated that 100,000 years were necessary for the terrestrial ecosystem to recover and 3 million years for the marine one.

Between April and May 2016, research drilling was carried out in the impact zone a few kilometers off the northwest coast of the Yucatan Peninsula in waters only 20 meters deep *(Fig. 2.25)*. It was Expedition 364 of the International Ocean Discovery Program. This study confirmed and refined previous data about the magnitude of the impact.

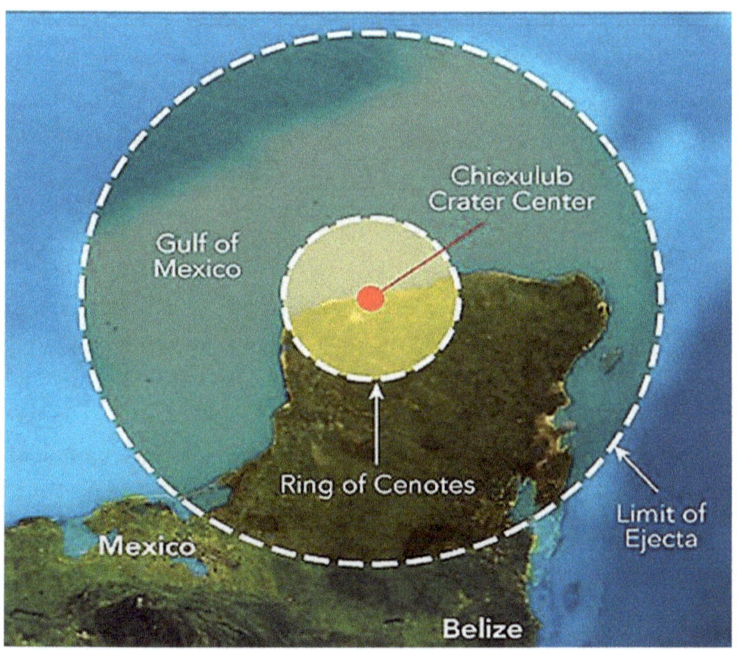

Figure 2.25 First report of expedition 364 in Chicxulub.
Credit: Mexican Academy of Sciences.

Chicxulub Crater is the only well-preserved ring-peaked crater on Earth (like the ones we see on the Moon). It was discovered that the impact caused a ~112 miles diameter crater with crust and mantle deformation up to 25 miles deep, melting the soil at the center of the impact more than 7.5 miles into the crust. The titanic shock originated sometime later, forming a vast hydrothermal system (underground rivers and cenotes). The crater eventually became an oasis for the recovery of marine and terrestrial life, allowing its expansion into momentarily empty niches that, when filled, defined a new local flora and fauna *(Kring, 2017)*.

Concerning this mass extinction event, it has been stated on multiple opportunities that the meteorite's impact point was probably in one of the worst places on the planet due to the characteristics of the ground where it impacted, and this was the direct cause of dinosaurs extinction. The meteorite's size was not significant, "a bee sting" compared to the diameter of the Earth, which is 12,742 km (7,917 mi). It is calculated that if the meteorite had impacted only 2 to 3 minutes before or after, the greater ocean depth would have caused more significant tsunamis to appear. The different characteristics of the seafloor would have formed a less toxic explosion cloud with lower gypsum content. It would have only caused the regional extinction of the dinosaurs but never global. If the meteor's passage had varied 10 minutes before or after, its trajectory would have been tangential to the Earth's surface. The bottom line with these variables is that humanity most likely would not exist.

One of the vertebrate animals that survived the cataclysm was mammals. They are pretty old since they appeared on Earth 225 million years ago, just about 6-7 million years after the first dinosaurs *(Alcober, 2010)*. They arose from a group of reptiles called synapsids that later evolved into therapsids more similar to current mammals and possibly with hair. Fossils of the first known mammal, *Adelobasileus cromptoni*, were found in the Tecovas Formation in West Texas, and it is the common ancestor of all mammals *(Lucas, 1993; Abdala, 2018)*.

The predominance and diversification of these animals occurred just after the impact due to the extinction of the dinosaurs that had kept them at bay; it also contributed to their quality of adaptation to feed on different sources and the homeothermic capacity that allowed them to have nocturnal habits. The evolution of therapsids over millions of years led to the emergence of four orders of mammals:[33] the *monotremes* such as the platypus and echidnas that are oviparous and have a cloaca, the *multituberculates*, called the "Mesozoic rodents," extinct 35 million years ago and named for the multiple cusps on their molars, the *marsupials*, for example, the current koalas and kangaroos and the *placentals* that emerged 90 million years ago and that make up most of today's mammals. There are more than 5,400 species nowadays. We humans, specifically, are descendants of the *Hominoidea* superfamily *(Rowe, 1988; Springer, 2011)*.

From this superfamily *(Fig. 2.26a and 2.26b)* arises the Hominid family and hence, the Hominini tribe in Africa from which seven genera descend, six of them (the oldest) now extinct *(Fig. 2.27)*. First, the *Sahelanthropus* lived 7 million years ago, when chimpanzees and humans diverged; in fact, this genus was the last common ancestor of both (*Homo and Pan*)[34].

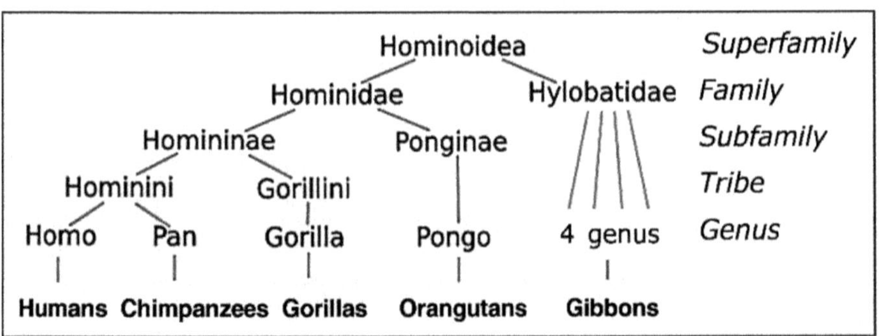

Figure 2.26a Hominoids Taxonomy.
Modified from: Petruss Hominoid Taxonomy (7.svg CC BY-SA 3.0).

---

[33] The geographic distribution of the origin of mammals is very complex. Still, one can find it in the magnificent article by Mark S. Springer et al. entitled -the historical biogeography of Mammalia. *Phil. Trans. R. Soc. B* (2011) 366, 2478-2502.

[34] The divergence in the chimpanzee and *Homo sapiens* genomic sequence occurred 6.5 million years ago *(Noonan, 2010)*.

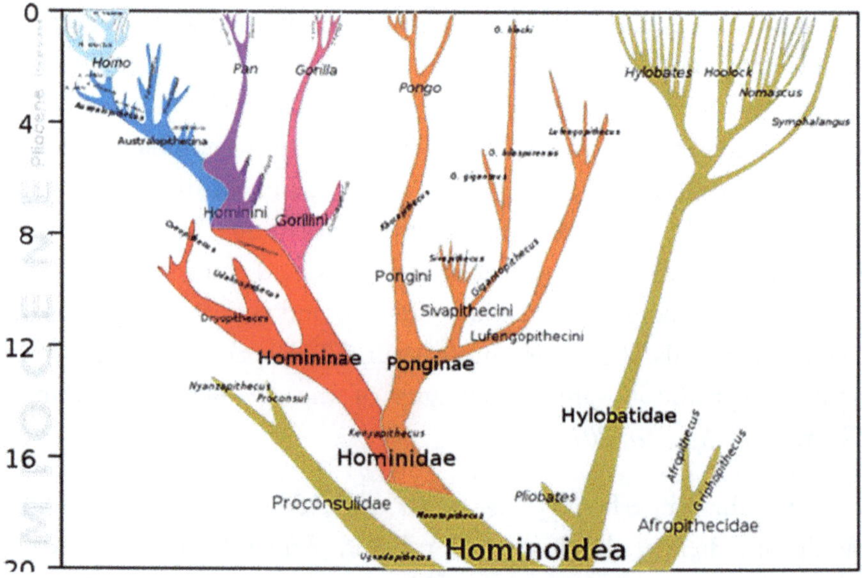

Figure 2.26b   Hominoidea phylogeny.
Credit: Dbachmann, CC BY-SA 4.0.

Later the Orrorins appeared six million years ago; they could stand upright[35]. Then, one and a half million years later, the Ardipithecus, with a 300 to 350 c.c. cranial capacity appeared.

Afterward, from four to two million years ago, *Australopithecus* existed in southern Africa that coexisted with three other genera which were more or less contemporary: the *Paranthropus*, also called robust australopithecines, which inhabited eastern and southern Africa, 2.7 years ago to 1.1 million years ago, the *Kenyanthropus* that existed from 4.1 to 2.9 million years ago and finally, the *Homos* our first ancestors that arose ~ 2.5 million years ago. Nobody knows with certainty from

---

[35]   Some biologists and anthropologists have concluded that the change in climate and vegetation with the progressive loss of the great forests in East Africa and the proliferation of grasslands (prairies and savannas) were decisive about 7 million years ago. The evolution of hominids to become modern humans accelerated because it forced our ancestors in an adaptive response to descend from the trees and develop the standing, whereby they had a wider field of vision to hunt and move quickly in the open area. In the same way, they also created new eating habits and social skills necessary for survival *(Klein, 2016: Uno, 2016).*

which species the first members of the genus *Homo* came; *Australopithecus africanus, A. afarensis ("Lucy"), A. garhi,* and especially *A. sediba* have been proposed, but there is no general agreement *(Wong, 2012).*

It has also been suggested that *Kenyanthropus platyops* could have been the ancestor of the first *Homo* that arose 2 to 3 million years ago and whose oldest representatives are *Homo habilis* and *rudolfensis* of 1.30 m in height and an average cranial capacity of 600 to 750 c.c. Precisely with the appearance of these first two species of humans, the Paleolithic Period, which means "ancient stone," begins and covers 99% of our genus's existence and ends approximately 10,000 years ago. From that moment on, the Neolithic or "new stone age" began and was completed in 4,000 BC., when the Metal Age started.

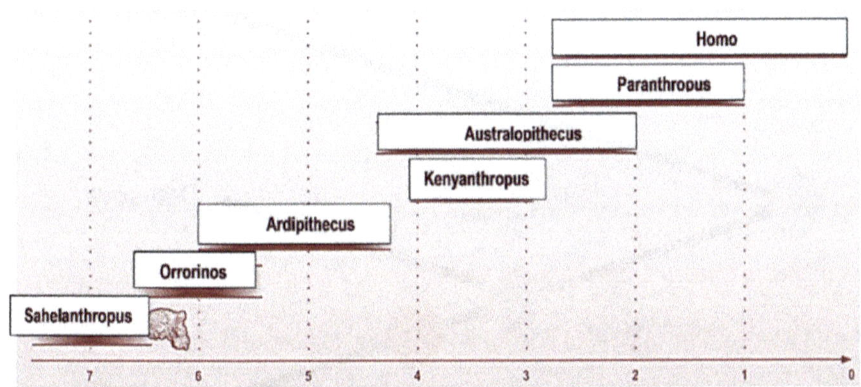

Figure 2.27 Chronology of the genera of the «Hominine Subtribe»
(Millions of years).
Modified from: Martin Sauer. CC BY-SA 3.0.

1.9 million years ago, another species evolved, *Homo ergaster,* the first to emigrate from Africa. It had an average cranial capacity of 850 c.c., from which *Homo erectus* descends (although they were contemporaries). It was recognized for a long time as the Man of Java, with a more significant stature (1.70 m = 5' 7") and cranial capacity that ranged from 850 cc. in the first specimens up to 1,100 c.c., when they disappeared ~ 113,000 years ago *(Rizal, 2020)*. 1.2 million years ago, *Homo antecessor* emerged in Europe, which had a weight of 90 kg and a height of 1.80 m (5' 11") with a cranial capacity between 1,000 and 1,150 c.c. and from which two species descended: the *Homo heidelbergensis*, whose remains discovered in Europe date back to 600,000 years, the members of this species had a cranial capacity between 1,100 and 1,400 c.c. and some of them, measured more than 1.80 meters in height and the *H. neanderthalensis* (descendants of a branch of *H. erectus*) of 1.70 m that lived in Europe and Asia 430,000 years ago and became extinct 28,000 years ago, the last being a group that he inhabited an area of the Gibraltar peninsula; these had a cranial capacity of 1,600 c.c., *(Avise, 2010; Paixão-Côrtes, 2012; Arsuaga, 2014)*.

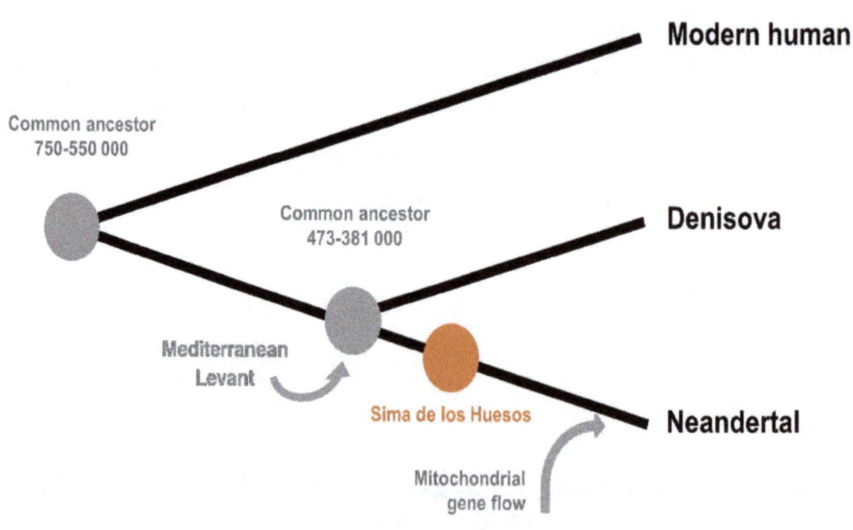

Figure 2.28 Phylogenetic tree of recent human lines

Modified from Meyer, M. (2016). *Nature* 531 (7595): 504-507, and Hershkovitz, I. (2021) *Science* 372 (6549): 1424-1428. UtaUtaNapishtim 233. Public domain.

Genomic analysis to find the chronological divergence between Humans and Neanderthals *(Fig. 2.28)* indicates that both species' most recent common ancestor is about 700,000 years. After that, the separation of the ancestral populations gave rise to modern humans and Neanderthals; it occurred 470,000 years ago *(Noonan, 2010; Meyer, 2016)*.

On the other hand, 700,000 years ago, a *Homo* genus appeared on the island of Flores in Indonesia. Because of that place, they have named *floresiensis* species. They were extremely short, only 1.10 m (3' 7") tall, around 25 kg (55 lb.) in weight, and with a cranial capacity of 420 c.c. These hobbits became extinct 12,000 years ago due to a volcanic eruption on their island, making them the most recent extinct *Homo* species. Currently, on the same island, there is a town called Rampasasa, where 77 families live, and 95% of the population measures less than 1.55 m (5' 1").

The sequence of evolution and migrations of the species of the genus *Homo* are complex and require coherence to locate us in time and place *(Fig. 2.29)*.

On October 28, 2019, a momentous event was published in *Nature* —the exact area where modern humans emerged 200,000 years ago was found. The region we come from is south of the Zambezi River in northern Botswana. This international study led by the Garvan Institute for Medical Research in Australia used chronological, ethnolinguistic, and geographic frequency distribution data across more than 1,000 mitochondrial genomes *(Chan, 2019)*. The region identified as the cradle of modern humans is a desert saline, but 200,000 years ago, it was Africa's most extensive lake system. This geographical area is more than 2,000 km southwest of Lake Victoria, where the origin of humanity was previously supposed to have taken place.

Nevertheless, experts question this study that claims to identify the birthplace of all humans. Instead, they say our species probably arose in many African places, not just in the Kalahari Desert or around Lake Victoria *(Gibbons, 2019)*.

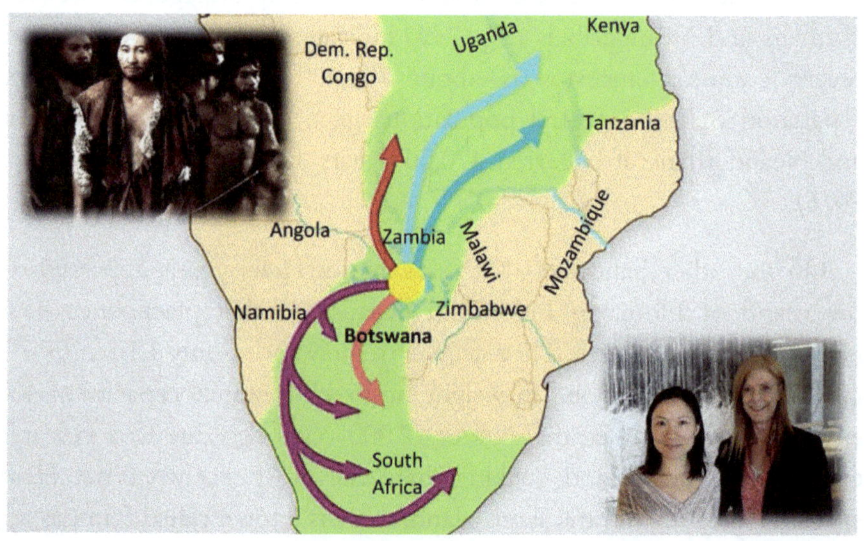

Figure 2.29   Cradle of Humanity. Northern Botswana.

Modified from Entressan/SPL. Ann Gibbons. *Science* doi:10.1126/science. aba0155.
Dr. Eva Chan & Professor Vanessa Hayes. Garvan Institute of Medical Research.
Chan, E. (2019) Human origins in a southern African paleo-wetland
and first migrations. *Nature*, 575:185-189.

Dr. Christopher Stringer, a British anthropologist from the University College of London, published in 2012 an article entitled *"What Makes a Modern Human."* He presents a scheme of millions of years with a coarse geographical location that is easy to understand *(Fig. 2.30)*.

- The contemporary *Homo floresiensis* of *H. habilis* originated in Africa from unknown ancestors (probably *Homo erectus*) ~ 1.75 million years ago. It migrated to the island of Flores 700,000 years ago, where it settled, becoming extinct 12,000 years ago *(Argue, 2017; Scardia, 2020)*.

- *Homo erectus* appeared 1.6 million years ago and spread from Africa to Western Asia, East Asia, and later to Indonesia. Its presence in Europe is uncertain, but it gave rise to *Homo antecessor*, found in Spain.

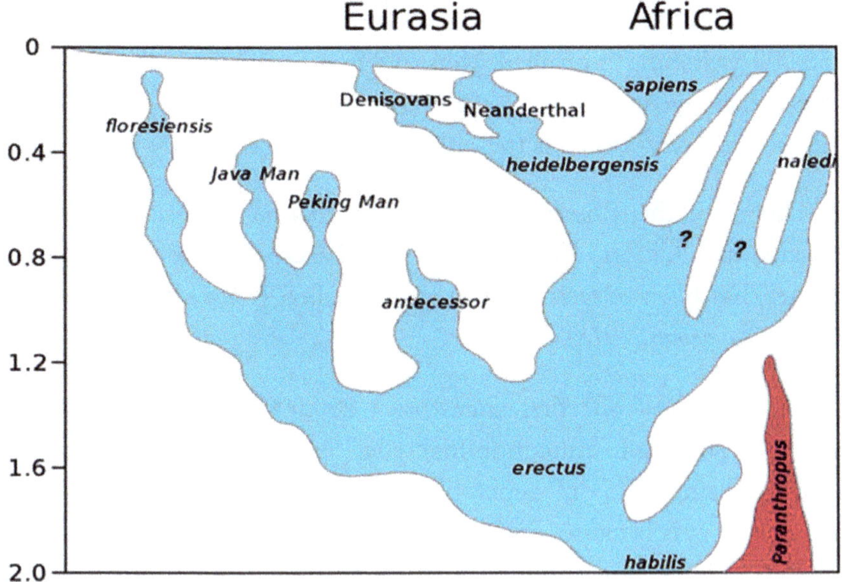

Figure 2.30   Genealogical tree of the Homo family in millions of years.

Modified from: Dbachmann Dec. 2017. Homo-Stammbaum. CC BY-SA 4.0.
Stringer, C. (2012). What makes a modern human. Nature, 485 (7396): 33-35.
Xu, D. et al. (2017). Mol Biol Evol. 34 (10): 2704–2715.

- *Homo heidelbergensis* originated from *Homo erectus* in an unknown location and spread throughout Africa, southern Asia, and southern Europe.

Homo sapiens emerged in Africa 195,000 ± 5,000 years ago and spread to Western Asia, Europe, and South Asia, finally reaching Australia and America.

- Early modern humans spread worldwide and interbreed with other descendants of *Homo heidelbergensis* —Neanderthals, Denisovans, and unknown archaic African hominins.

Man is a primate of the hominid family and the only current representative of the genus *Homo*. During the evolution of this genus, the most primitive species appeared and disappeared, leaving a genetic legacy among us. Finally, exclusively the sapiens species persisted with two subspecies: the paleo-subspecies *idaltu,* whose remains date back to 158,000 years and are called Men of Herto by the region of Ethiopia where they were found in 1997 (*Homo sapiens idaltu*), and ours, the subspecies *sapiens* which is the only one that persists *Homo sapiens sapiens.*

Therefore, the trinomial nomenclature that includes subspecies must remain. In addition, *Homo neanderthalis,* based on recent paleogenetic studies, has been reclassified and is now called *Homo sapiens neanderthalis (NCBI Taxonomy, 2018).*

The dating of our first ancestors ~200,000 years ago coordinates with the so-called "mitochondrial Eve" based on the mitochondria DNA transmitted from generation to generation exclusively through women. That is because, during the fertilization of the ovum, the sperm loses its tail and mitochondria, leaving only the mitochondrial genome of the ovum.

The mitochondrial Eve would correspond to the genomic information of a common female ancestor that shares the entire current population of human beings. This specific woman kept the DNA of this intracellular organelle in each generation because there were always women in her offspring; that is why it is transmitted until today. At the time of their existence, there were probably hundreds of primitive *Homo sapiens sapiens* women. However, this genetic information was automatically lost if there were exclusively male offspring during subsequent generations. That is precisely why "that Eva" is the genetic ancestor common to all humanity *(Relethford, 2008).*

In 2009, a molecular evolution dating technique was developed through DNA nucleotide sequences that date the divergence of two species. This technique is called the evolutionary or molecular clock. This technique estimates that this woman lived approximately 200,000 years ago; this information coordinates perfectly with the time when our oldest ancestors lived, the natives in the Botswana region, and later the men of Kibish.

In the same way, through this molecular clock, our "Y chromosomal Adam" has been determined to correspond to the patrilineal line of the only male ancestor from which the DNA of the Y chromosome of the entire male population of current *Homo sapiens sapiens* would converge.

The most recent information places this ancestor between 202,000 and 209,000 years ago; therefore, these genetic Adam and Eve were not precisely contemporaries *(Callaway, 2013; Méndez, 2013; Karmin, 2015).* Thus, mitochondrial Eve and chromosomal Adam are *"More Recent Common Ancestors."*

In 2010, paleo-anthropologists began sequencing the DNA of our ancestors in large populations so that the evolution of hominids can now be ratified or redefined according to the findings, for example, of the percentages of the Neanderthal genome or Denisova men in different communities worldwide. Now, population geneticists and biologists can work under a more solid base. Also, it allows them to explain, in addition to the physical characteristics of the inhabitants of different regions, the reason for the greater or lesser incidence of multiple primarily metabolic diseases whose origin starts from diverse common ancestors. Today is also possible to confirm the migrations that began in the south-central region of Africa.

In September 2015, a group of American paleontologists announced the discovery of a new species of *Homo* called *naledi*, never seen before. This discovery occurred in a region of South Africa 50 km (31 mi) northwest of Johannesburg in the Dinaledi chamber of the "Rising Star" cave. The remains of 15 bodies were found showing unique characteristics such as a minimal cranial capacity of 560 c.c. for men and 465 c.c. for women, very similar to the famous *"Lucy,"* *Australopithecus afarensis* that lived 3.2 million years ago in Ethiopia and measured one meter in height with 27 kg (60 lb.) in weight.

In addition, the *H. naledi* brain was tiny but complex, sharing structural similarities with the modern human brain *(Holloway, 2018).* In May 2017, the investigators found another chamber *(Chamber Lesedi)*

100 meters northwest of the Dinaledi chamber with 131 remains of the same hominids *(Hawks, 2017)*.

The first estimates about its antiquity based on morphological features placed it around 2 million years ago; however, *Homo naledi* is much more recent. It is located 236,000 to 335,000 years ago, in the middle of the Pleistocene (Dirks, 2017). They measured 1.50 meters (4' 11") with a weight of 45 kg (99 lb.), so the relationship between their body and cranial volume does not coincide with the observations of other contemporary *Homo* species, except *Homo floresiensis (Fig. 2.31)*. Still, the latter was 50 cm (1' 7") smaller.

On the other hand, *Homo naledi* had very modern hands and feet similar to ours; he walked upright *(Harcourt-Smith, 2015)*. However, he had the trunk, shoulders, pelvis, and the most primitive or australopithecine-like femurs *(Berger, 2015)*. They easily climbed trees.

He could handle wooden and stone tools *(Kivell, 2016)* and deposited his dead in specific places such as the Rising Star cave. This critical finding alleges our ability to associate certain hominin species with particular cultures and behaviors in the past. For example, before this discovery, it was thought that the only ones to bury their dead were *Homo neanderthalensis* and *Homo sapiens (Berger, 2015)*. It is equally disturbing to have managed to find a member of the genus *Homo* with features similar to *Australopithecus* and at the same time with modern anatomical characteristics. This finding formally contradicts the theory that the evolution of humans was relatively linear from an ancestral heidelbergensis-like morphology to that of *Homo sapiens (Gibbons, 2015)*.

In April 2019, the discovery of a new member of the *Homo* genus named *luzonensis* was announced. He was found in the Callao Cave of Luzon Island in the Philippines; he was 1.20 m tall, and his existence dates back between 50,000 to 67,000 years ago. It has several similarities with *Homo floresiensis (Figs. 2.31a and 2.31b)*, although this island is 3,000 km away from Flores Island *(Detroit, 2019)*. It is probably that the restriction of local food resources in both species ultimately reduced their body mass to have greater chances of survival. Therefore, *Homo erectus* is considered the common ancestor of both species.

Figure 2.31a - On the left: *a reproduction of Homo floresiensis physically similar to Homo luzonensis.*
On the right: *reconstruction of Homo naledi.*

Modified from: John Gurche / Mark Thiessen / National Geographic and Smithsonian National Museum of Natural History.

The imperfect chronological control over the African middle Pleistocene record does not allow an orderly progression through time from "archaic sapiens" to "modern sapiens." The origin of humans was multi-regional, especially austral and central in the African continent, with many potentially inter-fertile subdivisions of various evolving sapiens species. These processes result in the modern *Homo sapiens* —genetically, morphologically, and behaviorally, but « there was never a single center of origin » *(Stringer, 2016)*.

On the other hand, a fascinating fact, which was also a strong driver of evolution, was the use of fire. *Homo erectus* was already cooking its food more than 1.5 million years ago, and *Homo sapiens ate a diet rich in cooked food since its emergence.*[36] The use of fire for food processing and large amounts of meat in the human diet gradually reduced the time spent on food - from 48% each day required by early *Homos* to less than 5% in modern *Homo sapiens*. This fact has been proven by specific observations such as the reduction in the size of the molar pieces, the size of the jaw, and the length of the intestines. They could use the time they previously used to eat to think, create, and have other intellectual activities, which contributed to the increase in cranial capacity and rational intelligence *(Organ, 2011)*.

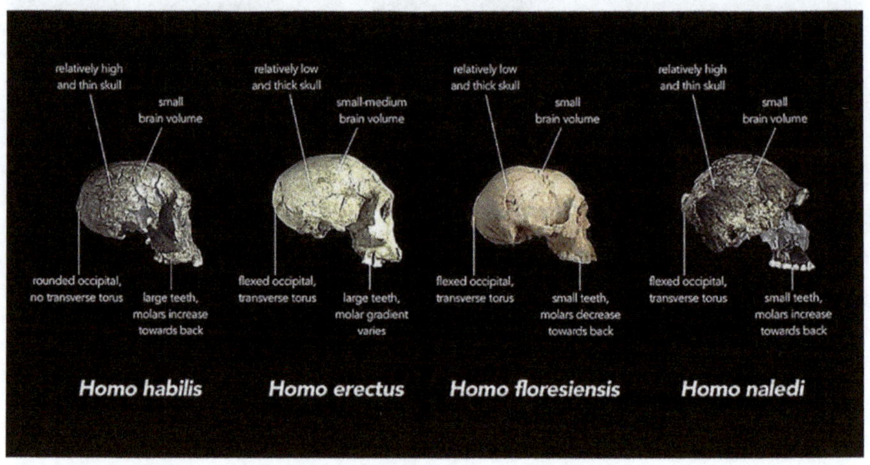

Figure 2.31b   Comparison of skull features of Homo naledi with another small-brained Homo.

Credit: Stringer, Chris (2015). "The many mysteries of *Homo naledi.*" eLife 4: e10627. DOI:10.7554/eLife.10627. - CC BY 4.0.

In the last 15 years, a solid basis has been identified that places nutrition, specifically docosahexaenoic acid (DHA), in a decisive role in the evolution of human intelligence. In fact, a turning point coincides with the inclusion of fish and shellfish in the diet. The

---

[36] In June 2022, the finding of the use of fire 1.8 million years ago in a region near Galilee, Israel, then occupied by early *Homo erectus*, was published in the journal PNAS *(Stepka, 2022).*

multigenerational exploitation of shellfish by coastal inhabitants coincided with the rapid expansion of gray matter in the cerebral cortex that characterizes the modern human brain. Late archaic humans (*Homo neanderthalensis*) were predominantly supplied with protein from the red meat of bears, wolves, large felids, and hyenas.

There was no evidence of freshwater aquatic species or marine sources of protein in the bone collagen of Neanderthal samples. Instead, shellfish consumption was a "staple" of the diet of early modern humans (Middle-Upper Paleolithic period). Depending on the geographic region, freshwater or marine protein sources constituted between 10 and 50% of the diet of early modern humans *(Bradbury, 2011)*.

There will be new and exciting discoveries in the future, such as in several regions of Africa, Asia, and the deposits in the Sierra de Atapuerca in Spain.[37] Meanwhile, advances in genomics have given a new direction to evolutionary biology allowing us to explore and specify molecular sequences over time.

Indeed, in 2014, a new science was inaugurated, *Epigenetic Paleoanthropology*, which, through the use of algorithms reconstruct the methylation patterns in ancient DNA samples, has made it possible for the first time to combine morphological information with epigenetics and thus be able to infer physiological traits of the human species *(Schneider, 2014)*.

In June 2021, in the city of Ramla in Israel, the remains of a man around 130,000 years old were discovered. That man was part of a group that has been called *Nesher Ramla Homo*. This human group, the "pre-Neanderthals," inhabited the Near East about 400,000 years ago. Now it is known that it was in the "Mediterranean Levant" -east of

---

[37] The archaeological site of Atapuerca is located in the province of Burgos in northern Spain. Several areas have found various remains of *H. antecessor* from 400,000 to 800,000 years old. In addition, more recent remains of *H. heidelbergensis* were also found with their tools. In this place, every 2 to 3 years, discoveries are made.

actual Spain- *(Fig. 2.28)* where *Homo Neanderthalis* emerged *(Hershkovitz, 2021)*.

Reconstruction of the origin and evolution of man through time has been challenging. The precise sequencing of our ancestors has been full of advances and setbacks, errors and corrections, and even frauds such as the famous *"Piltdown Man."* Considered the missing link for more than 40 years, "found" in 1912 in Sussex County in the south of England, which turned out to be a combination of remains of an orangutan skeleton, the tooth of a monkey, and the skull of an ancient human being.

The Future Human Being

Evolution from the beginning of life does not stop, and the physiological and physical characteristics that predominate in humanity today could be different in the future. It is surprising how we can notice even trivial changes such as the facial features of people in periods of only 100-200 years. For example, when we look at the old family photo album, how different the faces of great-grandparents and great-great-grandparents are compared to the faces of our couples and children. We say we see old faces; the same happens with the divas and gallants of the early twentieth-century cinema. Their characteristics and physical attractiveness differ from those of today's artists (although this is mainly due to clothing and grooming).

Interesting projective studies of humanity have been carried out based on historical-adaptive and sociocultural knowledge that allows us to take a glimpse of the functional and anatomical characteristics of the human being in the following generations. Let us see what we mean by this.

One of the most tempting aspects to mention is that our brain will be more prominent in the future, and therefore we will also be more brilliant. However, we must consider some factors necessary for this to occur. First, a biological impairment results from the relationship between the maternal pelvis's diameter and the fetal brain's growth. Second, laborious and difficult deliveries are significantly more frequent in humans than in other hominids because of the frequent disproportion between the size of the baby's head and its mother's pelvis. Third, the female pelvis has undergone several adaptive changes, which is very different from the male ones. Still, this adaptability has limits because standing is mandatory for humans and not optional like the other hominids *(Fig. 2.32)*.

Interestingly, it has been observed that mothers with large heads have larger pelvises since they will give birth to babies with large heads, which objectively indicates that the size of the head is genetically determined. Humans are born with approximately 66% of the head size present in the adult stage *(Underdown, 2016; Rosenberg, 2017)*.

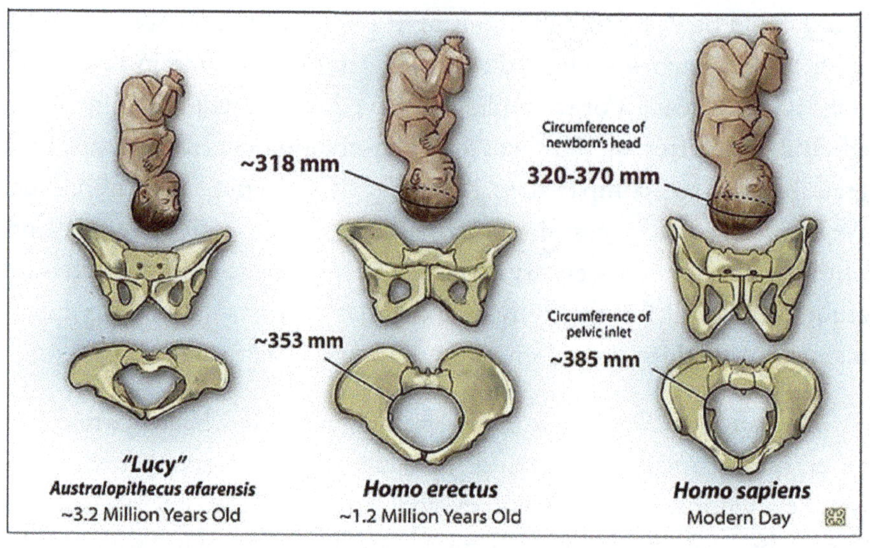

Figure 2.32 Evolution of the pelvis in hominids
Credit: Zina Deretsky, National Science Foundation. www.nsf.gov.

An evolutionary alternative could be that the term of human pregnancy, which is around 280 days, is reduced since the circumference of the fetal skull grows at a rate of ~0.6 cm per week at the end of it, which would allow babies to be born with a smaller cephalic circumference. However, this would imply many unlikely adaptations such as accelerated neurogenesis and the maturation of vital organs, such as the lungs.

However, the most important thing is that it has been observed that the opposite has happened evolutionarily. Within the genus *Homo*, over time, the head has become smaller. The Neanderthals had a higher cranial volume and a larger brain than modern humans have today *(Fig. 2.33)*. Not only Neanderthals but early Homo sapiens also had a larger skull and brain than the average we now have. The division of work activities in society and living in a civilization that reduces the risk of dying hunted by predators is why that large portion of the brain destined for survival has been lost because of self-domestication [38] *(Hood, 2014)*.

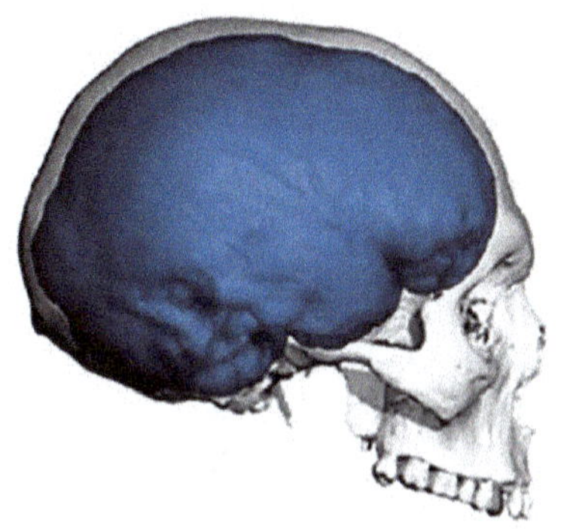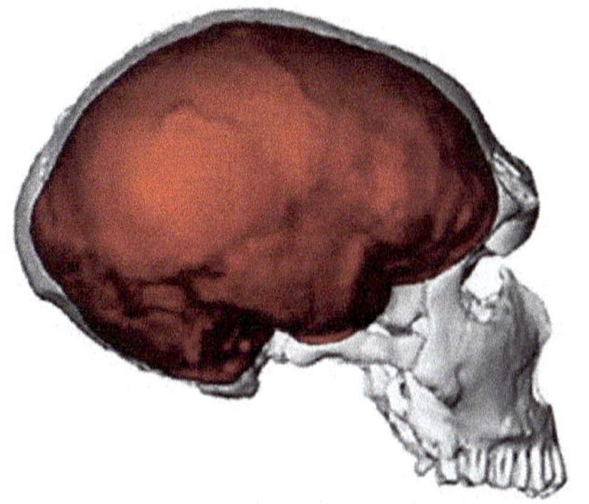

Figure 2.33 Differences in shape, cranial, and brain size between:
Left: *Modern adult human.* Right: *Adult Neanderthal.*

Modified from Hublin J. et al. (2015) Brain ontogeny
and life history in Pleistocene hominins.
Philos Trans R Soc Lond B Biol Sci., 370 (1663).

On the other hand, the high encephalization among hominids began 2 million years ago, just over half a million years after the genus *Homo* was established. Moreover, the human genome has redundancy in genes

---

[38] This same phenomenon is observed in domestic animals compared to their wild counterparts. For example, wolves have a larger brain than domesticated dogs of the same body size.

that generate neural development, and paleogenetic studies have shown that many new genes appeared in recent evolution. For example, 1,228 neo-genes were incorporated in the hominoid stage, of which 389 are unique to humans. Surprisingly, 198 hominoid and 54 human genes have their expression directed towards the prefrontal cortex, the most recently evolved brain structure related to advanced cognitive functions *(Chen, 2013; Hublin, 2015)*.

These facts allow us to reach two conclusions: first, intelligence is not necessarily directly proportional to the size of the brain; if so, whales and elephants would be more intelligent than human beings; and second, evolutionary adaptability is responsible for the increase in the complexity of the central nervous system and specifically of the neocortex in humans. Thus, a sophistication of higher brain functions occurred primarily through the increase in the number of specialized neurons and the multiplication of their interconnections. It ultimately results in a more complex connectome, resulting in humans having the highest degree of intelligence in the animal kingdom.

In the future, human beings will not have a significantly larger brains. Instead, modeling and regulating their nervous system to adapt to the new needs will continue. This process is carried out, among other actions, through the so-called « axonal and synaptic pruning, » a mechanism by which exuberant branches of axons and dendrites are eliminated, and their synaptic junctions are eliminated. This mechanism allows the refinement of neural circuits and facilitates plasticity and neuronal activity mediated by experience. The process occurs in two stages of life: the first during the two years after birth and the second during adolescence *(Riccomagno, 2015)*. As expected, these adaptive prunings have not been the same throughout history due to the difference in specific needs that humans have according to their environment. For example, compare the man who lived 10,000 years ago with those who lived in the 16th century or will live in the 22nd century.

Nevertheless, the variations have occurred historically. Each human being has its particular pruning process, partly due to its genetic load, the stimulation it receives during its first years, and later depending on its specific interests during adolescence. In this way, a boy's brain who prefers art will gradually become different from another who likes mathematics or biology, and thus each one will develop specific skills.

Ian Tattersall, the paleoanthropologist and curator of the American Museum of Natural History in New York, affirms that the evolution of man due to globalization will practically come to a halt. His theory states that when a genetic mutation occurs in isolated populations, it is stably transmitted to the following generations because it offers some benefit. Eventually, this mutation becomes the norm; but if the population is not dispersed, crossing multiple human settlements makes potentially significant mutations much less likely to establish themselves in a local genetic pool.

The genetic inertia in a large population is such that it resists substantial change, and that is what happens in current circumstances with the billions of inhabitants in the world migrating from one country to another with the consequent multiethnic interbreeding.

So based on this, any significant evolutionary novelty for humans is doubtful. Even as a consequence of this thesis, precisely the opposite will happen. The high genetic crossing between the different ethnic groups will make human beings more similar in the future, and the phenotypic-racial differences will be less evident.

Another exciting aspect arises from one of the bases of Darwin's evolution, which refers to the survival of the fittest; however, the rule nowadays, with the advances in medical science, is that most of us remain alive for many years. This data means that the survival of the fittest or strongest as a concept of evolution has disappeared for most human beings. This thesis by Welsh geneticist Stephen Jones joins Tattersall's because a significant evolutionary trait in future humanity is highly unlikely *(Owen, 2009)*.

However, another group of researchers observes evolutionary phenomena happening today, which could continue in the future. For example, given that women of shorter height and some overweight tend to have more descendants than their peers of different physical constitutions, most women in the future could have these characteristics *(Byars, 2010)*. On the other hand, attractiveness between couples could be realized in the future primarily by intelligence rather than physical appearance.

As technology becomes more complex, humans will need more intelligence to master it, and that trait could guarantee economic success. It is also likely that obesity will continue to be a health problem for some time and associated with a decrease in muscle strength compared to previous generations; However, the exact opposite could happen, and employing artificial selection, when this becomes the main protagonist, it will allow future parents to have the best children that technology can give them: *"Citius, Altius, Fortius,"* faster, taller, more muscular and of course, brighter.

One of the fears that could become a reality in a few decades is that there could be two types of human beings: those conceived through genetic engineering, "custom-designed babies," and those created "naturally." That would eventually cause a disproportion in performance capabilities, job opportunities, and income, translating it into success in life. In this way, it could come true with some variants, the novel of the British writer Aldous Huxley, *"Brave New World,"* and Globalized Eugenics could succeed before the end of the 21st century.

Finally, we would have to add that, perhaps in the very distant future, probably centuries, natural selection could once again act in human beings as obviously as it did with certain dispersed human groups with independent evolutions. Thus, for example, in the regions of what is now Rwanda and the Congo, for centuries, there have coexisted tribes of Watusis or Tutsis that have an average height of 1.96 m (6' 5") for men, and 1.76 m (5' 9") for women with the Pygmy peoples whose average height for men and women is less than 1.50 m (4'11").

Precisely, these examples could occur in the future, not in our world, but in human colonies on other planets, which, due to the different natural environments, could cause the generations that decide to occupy these habitats to mutate to achieve gradual optimal adaptation to face the challenges that these new environments will pose for them.

As a final point before concluding this section, this is a thought from Dr. Mihály Csíkszentmihályi, Dean of the Psychology Department at the University of Chicago and head of the Sociology and Anthropology Department at Lake Forest University:

"Evolution has been advancing towards the peak of complexity and, whether we like it or not, the acme at the moment is us. It is up to us that evolution continues to produce more complex shapes in the future. So, we can help make this world a more incredible place than ever or accelerate its return to inorganic dust."

Consciousness

*" When the woman saw that the tree's fruit
was good for food and pleasing to the
eye and desirable for gaining wisdom.
She took some and ate it, and she also
gave some to her husband, who was
with her, and he ate it.*

*Then the eyes of both of them were
opened, and they realized they
were naked, so they sewed fig
leaves together and made
loincloths for themselves."*

Genesis, 3:6-7

Introduction

In this final chapter, we will learn about consciousness and its transcendence as a culminating point in the evolution of the Cosmos. We have examined the impressive series of events that happened over millions of years for life to appear and thrive on our planet. However, now we will concentrate on the meaning of consciousness, through which the Universe undoubtedly perceives its existence.

The Merriam-Webster Collegiate Dictionary, probably the most influential and respected English-speaking dictionary, defines consciousness as:

1) The quality or state of being aware, especially of something within oneself.

2) The state or fact of being conscious of an external object.

3) The state of being characterized by sensation, emotion, volition, and thought.

4) The upper level of the person's mental life is aware of contrasts with unconscious processes.

5) Knowledge of the social and political environment.

"Consciousness" is the function of the human mind that receives and processes information directly linked to the five senses, reasoning, memory, imagination, and emotion. On the other hand, *"Conscience"* is the inherent ability to perceive what is right and wrong, and on the strength of this perception, control, monitor, evaluate, and execute actions *(Vithoulkas, 2014).*

There are differences between *awareness* and *consciousness*. Although they seem to have the same meaning, it is not. Awareness is knowing something that exists somewhere. On the other hand, consciousness is the state of being aware of something. Consciousness is the quality or the state of awareness. Awareness is a necessary condition of consciousness.

In *Figure 3.1*, we can observe that vigilance (as a synonym for conscious or awake) and awareness both are interdependent, but they are not always directly proportional; for example, people with Alzheimer's may have an excellent level of wakefulness with an inferior content of consciousness *(Boly, 2013)*.

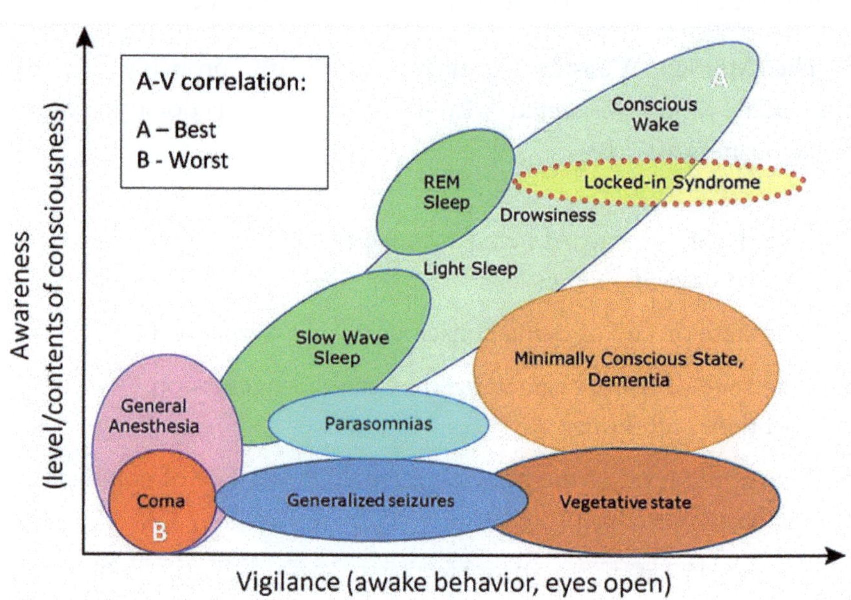

Figure 3.1 *Vigilance state associated with the level and content of consciousness.*

Modified from Boly, M. et al. (2013). Consciousness in humans and non-human animals: recent advances and future directions. *Front Psychol,* 4: 625. doi: 10.3389 / fpsyg.2013.00625.

On the other hand, Locked-in Syndrome is a conserved state of wakefulness and consciousness with quadriplegia and lower cranial nerve palsy leading to the inability to move, speak, or communicate except through eye movements. It is caused by infarction, hemorrhage, or tumors of the annular pons of the brainstem *(Maiese, 2019)*.

Consciousness is not an exclusive ability or attribute of human beings. Instead, consciousness is present in all living beings in different magnitudes based on their ability to perceive themselves and their environment.

Historically, the leading disciplines in the knowledge of this subject were philosophy and psychology, later joined by cognitive science and neuroscience. The latter, although biologically based, is closely related to all the others and even to quantum physics. Thus, it is only through all these disciplines that consciousness can be studied as a whole. We will start with the biological and evolutionary bases.

Biological Basis of Consciousness

The evolutionary biology of consciousness starts from the most basic level, the sensitive-perceptual in cells, to the highest, the sense of self or self-consciousness that is the main constituent of higher consciousness. It can be identified through its three components: *behavioral, neuronal, and informational.*

The *"behavioral"* components include those related to high cognitive abilities, such as behavioral flexibility, verbal skills, memory, theory of mind (the ability to attribute thoughts and intentions to other people or animals), object constancy, deductive ability, and multistable perception (in the face of ambiguous images). It has been proven that they exist in humans, birds, and primates.

The *"neuronal"* components encompass the thalamic-cortical system for mammals, including humans, and similar circuits exist in other non-mammalian animals such as birds *(Fig. 3.2)*. A transcendental fact for us humans concerning higher consciousness is the recent discovery about the structure where my self is located, that entity inside my head where I illusively feel that my being is found. This structure is called the hippocampus, located in the internal and central area of the temporal lobes below the cerebral cortex *(Guterstam, 2015)*.

Finally, the *"informative"* components are the members of the sensorimotor consciousness, which allows us to respond using bodily movements to the various stimuli perceived through the sense organs. These information components are the basis for integrating the sense of ourselves and separating what belongs to us from what does not. The other informational elements of the sensorium, such as colors and shapes, are dependent on the visual system; the sounds, music, and noises of the auditory system, flavors and odors of the taste and smell systems, and finally, the perception of textures, pressure, temperature, and identification of objects through touch. By integrating the three components, the degree of consciousness present in a particular species is defined as a biologically generated phenomenon that requires at least a population of cells that can detect sensory stimuli, such as light, sound, or chemicals, and the capacity to communicate with each other, which is a set of neurons within a multicellular organism.

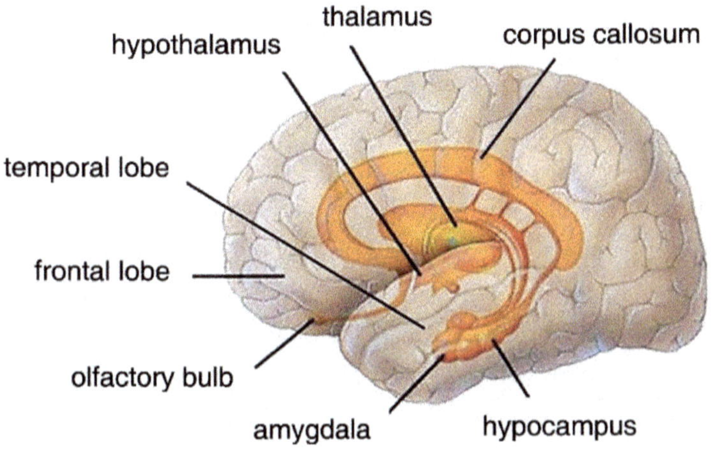

Figure 3.2 Anatomy of the brain. Neural components of consciousness.

Modified from: Pioneering Minds. Limbic System.

There must be an excellent balance between its excitatory and inhibitory connections for stable multicellular neuronal functioning[39].

---

[39] The release of specific neurotransmitters determines the excitatory and inhibitory activity of neurons. Glutamate and aspartate (excitatory) and gamma-aminobutyric acid "GABA," and glycine (inhibitory). However, it is essential to remember that communication between two neurons is not only carried out by neurotransmitters, but there are electrical synapses in which the current is generated by ionic action

On the other hand, we have made significant progress in understanding the functions of neurons at the molecular level, for example, the role of sodium, potassium, chlorine, and calcium ion channels in cell membranes, as well as the intracellular mechanisms related to neurons, and neurotransmitter systems in the generation of consciousness. However, we do not know how neurons generate the representation of reality from these elements. This question is one of modern neuroscience's most significant challenges in understanding ourselves *(Butler, 2012)*.

Although much remains to be understood, we have learned more about the brain and its functions in the last 50 years than what has been achieved since *Homo sapiens* emerged.

Advances in technology, such as imaging, have opened an immense door that leads us to discover the deep mysteries of the most complex organ that nature has created. Every day we are more familiar with names like Tomography —axial computed, computed by positron emission, computed by emission of a single photon, and computed with ultrasound. In addition, Resonance Imaging —magnetic, functional, and diffusion magnetic; and the most recent High-Density Diffuse Optical Tomography (HD-DOT) *(Frijia, 2021)*.

The latter one is, among other objectives, the basis for the development of the Human Connectome Project (from newborns to the elderly). That will allow the precise mapping of the connections and the organization of neural groups of ~ 86 billion neurons that make up the human brain *(Azevedo, 2009)*, knowing that each neuron communicates with at least a thousand others.

But, not only imaging technology has allowed advances in neurophysiology and neuro diagnosis. Also, for example, the old electroencephalogram technique relies on quantitative electro-encephalography and magnetoencephalography. Both of them with

---

potentials that flow directly from the presynaptic neurons to the postsynaptic without the need of specific biochemical receptors.

different capabilities that complement its predecessor. With Spectroscopy Close to Infrared for functional neuroimaging and brain-computer interface, we can carry out a structural and functional mapping of the brain mass and, thus, better understand the physiology of our brain. That, in the end, is nothing more than *"the brain trying to know itself."*

Technology is facilitating neuroscience to take the leadership that philosophy had for many centuries in the knowledge of consciousness *(Chalmers, 2015)*. Regarding this statement, in the first chapter of *The Grand Design* (2010), Stephen Hawking and Leonard Mlodinow write: *"How can we understand the world in which we find ourselves?" What is the nature of reality?* Where does everything around us come from? Traditionally those are questions for philosophy, but philosophy is dead. This is because philosophy has not kept up with modern developments in science. Scientists have become the bearers of the torch of discovery in our quest for knowledge. In this regard, we believe that philosophy's essence has historically been more related to questions than answers.

On the other hand, the current advances in theoretical neurobiology and systems in neurosciences are the starting point for developing the Theory of Understanding, which is one of the best ways to approach the nature of consciousness. The theory relies on the basis that cognitive and understanding capacities depend on groups of associated neurons that together form models that allow the perception of objects and sensations and thus respond adaptively mainly for survival purposes.

Evolutionarily living beings go from feeling to perceiving through these neural packages and then to learning, creating a primary memory [40]. Once learned, the next step is to understand, which leads us to

---

[40] Memory is a function that allows you to encode, store and retrieve information from the past. It is intimately linked to learning, but to some extent it is independent of understanding. Memory implies the retention of acquired information, whether understood or not. Instead, for understanding, comprehension of information is essential with or without complete memorization.

have the ability to adapt and innovate responses that are the most appropriate according to the stimulus that gave rise to them and even the power to predict external phenomena to adapt reactions and behavior *(Fig. 3.3)*.

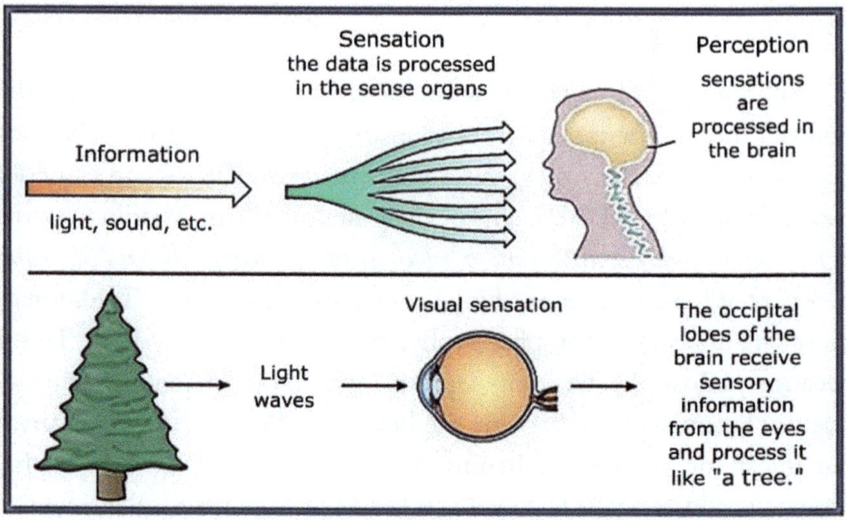

Figure 3.3 Information, Sensation, and Perception.
Public Domain.

Intelligence defines the ability of a living organism to vary and choose its adaptive responses to external conditions or stimuli in a way that ensures its survival or is successful in its purpose *(Trewavas, 2016)*. *Conscience* arising from the understanding suppresses the inertia to carry out the same learned behaviors in different conditions, creating new responses that are more appropriate to new challenges and that can adapt and anticipate environmental fluctuations *(Yufik, 2016)*.

In this way, we understand why consciousness and intelligence are closely related and are generally directly proportional. Both bases are sensation and perception, followed by learning and understanding. Consciousness is responsible for showing the present scenarios through the perception of the external and internal environment in the most attached way possible to reality according to the capacity of that organism to understand that reality. In such a way, the intelligence can choose the most appropriate responses to survive and achieve its goal. In turn, intelligence takes relevant information to consciousness, allowing it to understand better its internal and external environments *(Fig. 3.19)*. A simple way to conceptualize consciousness would *be "the use of mental images to represent reality (cause) and regulate behavior accordingly (effect)."*

The ability to have a comprehensive understanding is a unique quality of the higher species. In the case of a primitive organism, which does not have a nervous system as such, it can only detect, for example, the physical conditions or stimuli that occur on its surface, generate location commands towards the origin of stimuli, and have specific responsiveness and learning.

Such is the case of *Physarum polycephalum*, a single multinucleated cell. This protozoan can learn and anticipate simple repetitive signals, observed when stimulated by electrical impulses on its surface separated with equal time intervals. This cell responds by momentarily reducing its growth rate in each electrical stimulus during the experiment. Remarkably, after three identical stimuli were given at the same time interval, the same growth reduction response appears to a supposed fourth stimulus that was expected but no longer applied. It is the simplest model of perception and biological learning *(Saigusa, 2008)*.

Understanding is a component of intelligence (from the Latin *intelligentia* - ability to understand). It denotes the ability to adapt and manipulate behavioral models to various stimuli. As we have already said, intelligence defines the ability of a living organism to choose and vary its responses. Learning is also a component of intelligence involving memory and subsequent execution of stimulus-response associations. Therefore, it is the change in the behavior of an organism produced by a specific experience.

However, as we already mentioned, comprehension or understanding surpasses learning because it allows the creation of new and unknown responses; in this way, understanding is a component and a form of intelligence. Learning without understanding means limiting the number of answers because they are confined to past exposures. Understanding extends learning unlimitedly, allowing remodeling, simulation of the future, and anticipation of effects according to causes. Understanding is a component of intelligence and the awareness of the outer and inner world. *(Trewavas, 2011).*

Another biological and evolutionary theory about the origin of consciousness starts from *"Selective Attention,"* which basically involves choosing and increasing a particular signal among several present signs. The chosen one will be the most pressing and be related to survival *(Fig. 3.4).* This selective attention first appeared around the Cambrian explosion, about 541 million years ago, shortly after the evolution of the jellyfish. From this chronology, primitive consciousness appeared based on the advanced complexity of the nervous system presented by anamniotes, which would give rise to fish and amphibians later *(Feinberg, 2013).*

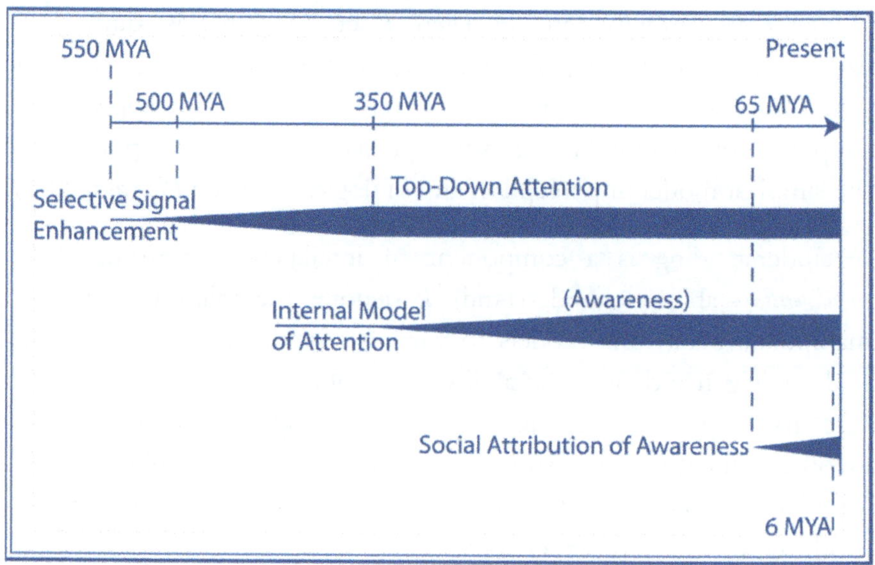

Figure 3.4 Evolutionary Neurobiology. From selective signal enhancement to awareness. (MYA = Million Years Ago).

Modified from Graziano M. (2014) Speculations on the Evolution of Awareness. *J Cogn Neurosci*; 26 (6): 1300-4.

Subsequently, 350 million years ago, information processing strategies of the *"top-down"* type appeared to control attention. It means the analytical and deductive processing of information that is carried out as a perceptual summary of the system by breaking it down into its parts and integrating it into a whole from which *"goal-based planning"* will emerge for survival purposes *(Fig. 3.5)*. Thus, the organism better understands its environment and has better adaptive responses to its needs.

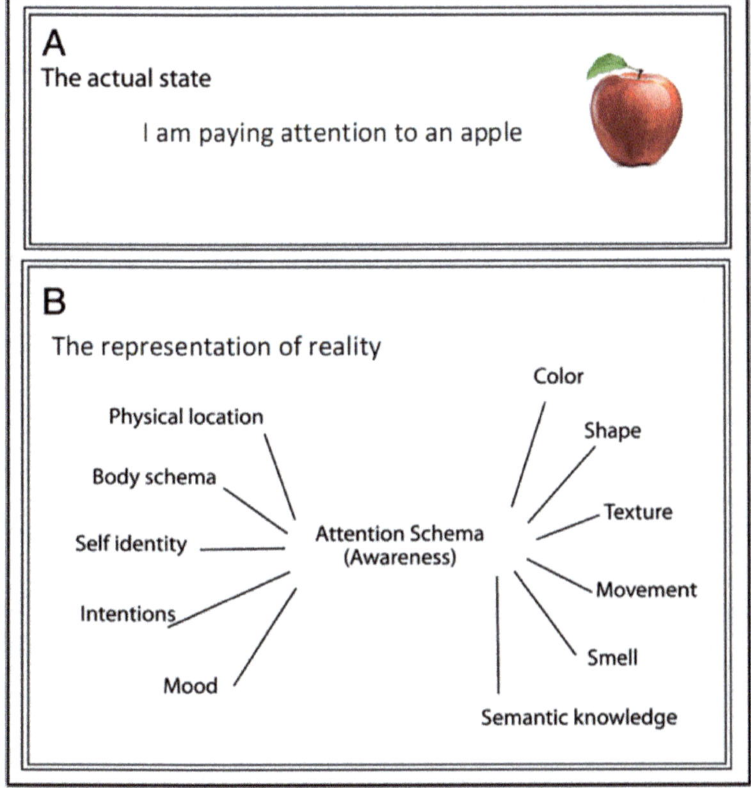

Figure 3.5   Information processing through top-down control attention.

Modified from Graziano, M. (2014) Speculations on the Evolution of Awareness. *J Cogn Neurosci*, 26 (6), 1300-1304.

Later, one hundred to two hundred million years of evolution were necessary to develop an internal model of attention that allowed the nervous system for the first time to attribute itself to the fact that it constituted a mind. Thus, the mind is understood as cognitive capacities such as perception, thought, memory, etc., so it was a possessor of knowledge with the ability to observe, predict, control, and hence can attribute consciousness to other beings *(Katsuki, 2014)*.

This social attribution of consciousness was first observed over 66 million years ago with the most intelligent dinosaurs, such as the Troodon (*Stenonychosaurus inequalis*). Troodon was a predatory dinosaur that was 1.20 m (3' 11") tall and 70 kg (155 lb.) in weight. But later, it was also present in some survivors of the Cretaceous-Paleogene mass extinction. Most birds and mammals have this well-developed ability, but it is at its best in humans.

Homo sapiens, from its origin, has been attributed consciousness to all its surroundings: to the rocks, to the rivers, to the stars, to the universe, and since then, it has created a spiritual world around it with deities, ghosts, and souls with a remarkable imaginative congruence creating cultures, religions, myths, and legends *(Graziano, 2014)*. The animistic play of children with objects and toys is undoubtedly an excellent example of attribution of consciences and symbolic thinking, which indicates that this characteristic is inherent to the human being and manifests itself from the first years of life *(Piaget, 1926 & 1945)*.

On July 7, 2012, an important conference in memory of Francis Crick entitled *"Consciousness in humans and non-human animals"* was held in Cambridge with representatives of the most prestigious Universities of the United Kingdom, Germany, United States, Japan, Australia, and Belgium. At this conference, each representative reported their investigations of the neurological correlation with consciousness, from fruit flies to hominids, including octopuses, parrots, and dolphins. Emphasis was placed on the ability of certain species to be able to recognize themselves in the mirror. It was long considered a uniquely human attribute, but now we know we share it with great apes, dolphins, elephants, magpies, and horses. This property translates to having the capacity for self-awareness, that is, to possess a higher consciousness *(Low, 2012; Baragli, 2021)*.

The coordinated collective activity of nerve cells (neurons and glial cells) is essential to ensure adequate ability to allow an organism to respond to the changing environment. In this context, consciousness is the global property of dynamic brain activity. However, an important question is how to describe this collective activity's organizing principles that allow characteristics associated with consciousness to emerge *(Mateos, 2017)*.

The biological basis of consciousness is located in the connectivity or coupling between neuronal groups whose link is more intense than other groups. The transmission of information between neurons takes place through electrical and chemical synapses. At electrical synapses, transmission occurs by sending positively charged ions from the presynaptic neuron to the postsynaptic neuron.

The electrical synaptic space between them, called the union gap, is less than 4 nanometers. It is there where the action potentials propagate from one neuron to another. These synapses are extremely fast and present mainly in the spinal cord and brainstem survival reflexes.

As for chemical synapses, there are biological junctions through which signals from neurons can exchange with each other and with non-neuronal cells, such as muscles or endocrine glands. Chemical synapses allow neurons to form circuits within the central nervous system and are essential for the biological functions that underlie perception and thought.

The thalamic-cortical circuits are the main pathways because they represent a more extensive and complex structure in humans than other mammals, contributing to humanity's exceptional cognitive abilities. These intercommunications are so broad that they are called thalamic-cortical radiations *(Fig. 3.6)*. They are mainly glutamate-dependent excitatory with inhibitory regulations through GABA (gamma-aminobutyric acid). All sensory stimuli that reach the brain, except smell, previously pass through the thalamus, and the three principal radiations are sensory, optical, and acoustic *(Izhikevich, 2008)*.

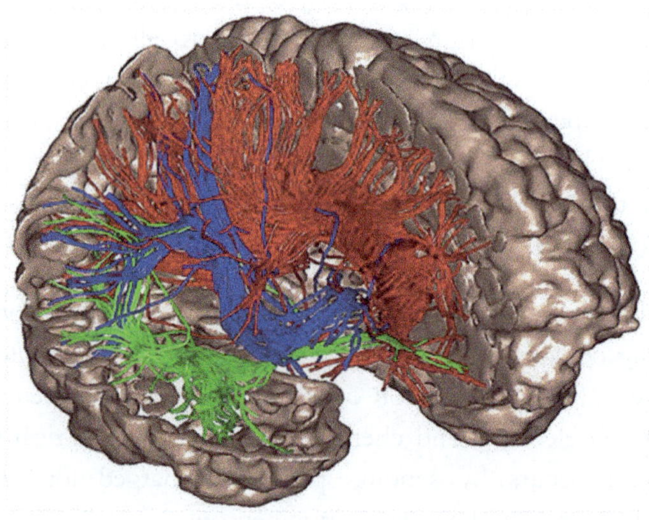

Figure 3.6 **Tractography of the white matter fibers of the thalamic-cortical system in the human brain (thalamic-cortical radiation).** Image obtained by diffusion tensor magnetic resonance.

Modified from Izhikevich E. et al. (2008) Large-scale model of mammalian thalamocortical systems. *PNAS* 105 (9): 3593-8.

The thalamus is the gateway to all sensory input, the portal and functional filter of sensations where perception occurs. Information is carried through significant pathways to the cerebral cortex, where the phenomena of cognition and consciousness are created. The thalamus sends and receives signals from all the lobes and convolutions of the cerebral cortex *(Min, 2010),* making a synchronized circuit that gives rise to all higher mental functions.

Recent findings show that human beings' *Neural Correlation of Consciousness* (CNC) is restricted to a privileged temporal-parietal-occipital zone with additional contributions from some anterior regions. Especially the so-called hot posterior cortical site includes sensory areas and the frontoparietal network involved in monitoring and information tasks *(Koch, 2016; Boly, 2017).*

The vast cortical areas related to consciousness (ascending reticular activating system) are associated with the content. However, those of the latter are located in more specific areas, as we can see in the lower part of *Figure 3.7.* Analyzing this approach, we must clarify that the Neural Correlation of Consciousness focuses on the contents of particular conscious experiences rather than on the actual capacity of the total subjective experience. That is an almost purely anthropocentric concept and, therefore, not applicable to most animals and plants *(Barron, 2016).* Later we will talk in detail about consciousness in non-human living beings.

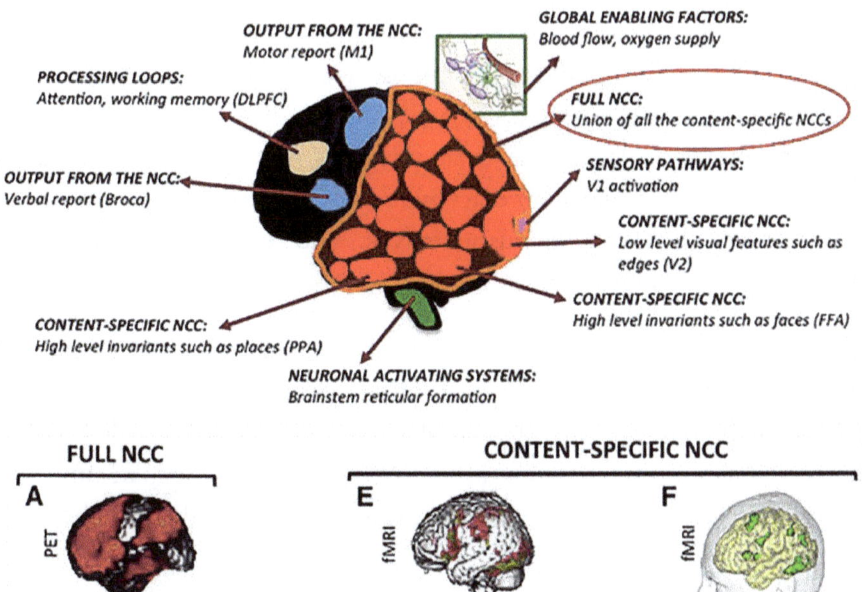

Figure 3.7 **Above: *Neural correlation with consciousness (NCC).* Bottom: *cortical regions related to the level and content of consciousness.***

Modified from Boly et al. (2017) Are the Neural Correlates of Consciousness in the Front or the Back of the Cerebral Cortex? Clinical and Neuroimaging Evidence *J Neurosci,* 37 (40), 9603-9613.

Consciousness is undoubtedly the great triumph of life. For us humans, it is that neural process through which we become aware of ourselves and our environment in time and space. It implies selective attention, multiple perceptions, the ability to carry out simple and complex actions, handle abstract ideas, have expectations or predictability, and treasure ethical and aesthetic values. Of course, consciousness has individual variations because it depends on unique genetic combinations and personal adaptive experiences resulting from learning obtained over time. Still, evolutionarily the level of consciousness is directly proportional to the complexity of the central nervous system.

Consciousness is a subjective phenomenon, accessible only to the individual who experiences it. It is unobservable and therefore not measurable, while brain activity is clearly objective and can be measured and studied. It is possible to demonstrate neural activity correlated with conscious experience, but we do not know how the first generates the second. In essence, colors do not exist in the external world, nor do sounds, smells, or tastes. Our brain develops these subjective elements as tools that allow us to interact with the outer world, transforming electromagnetic waves, acoustic waves, and odoriferous and taste chemical particles into perceptions.

It does not matter how the reality of the external world is because we only observe it through the filters of our sense organs in a fragmented way and analyzed in different regions of the brain, but perceived as a unified, coherent, and contextualized entity in time and space. In parallel, the inner world is captured and modified by the filter that conditions the emotions and physical and mental health state. So, consciousness as an experience of reality is exclusive to those who experience it, although it can be shared. Therefore, it is never possible to transmit it to others in its true essence and magnitude. Instead, our resources are verbal and body language, writing, painting, sculpture, music, smells, flavors, and textures, that is, sensory elements of communication.

Thus, conscious experiences are undoubtedly personal and non-transferable. The phrase (somewhat corny) *"the beautiful red fruit, with a mildly sweet taste, has floral aromas"* is an example of the personal reality that is not transferable as an experience, only as a description. Let us analyze the statement: beautiful is a personal-social concept that originates from that sensory experience derived from the perception of *"the harmony of forms in balance that gives us pleasure or satisfaction."* This experience is necessarily individual and contextual. Fruit is an element from the flowers that contain the seeds that will allow the plant's reproduction that gave rise to it. However, in the sentence, "the fruit" is an imaginative experience dependent on learning; for someone else, it will represent an apple, a plum, or perhaps a cherry. The word red denotes something that refracts light with wavelengths between 625 and 740 nanometers equivalent to 480-405 terahertz. Still, the experience of perceiving the "essence" of the red color, its tonality, hue, and intensity are uniquely individual. In the same way, the *"softness, sweetness and floral aroma";* are personal conscious experiences.

With this example, we arrive at the most challenging problem of consciousness *(Chalmers, 1995),* which is called «qualia » *Qualia* is the essence of things and their characteristics: the bluish of the blue, the brilliance of light. How can we convey to a man born blind the meaning of the tonality of colors? How could we inform a congenitally deaf person what a bird's chirp sounds like? Precisely what we cannot do would be *qualia,* the perceptual experience, and how we know the nature of things and their qualities. There is nothing that we know more intimately than conscious experience, but at the same time, there is nothing more difficult to explain. The nature of consciousness is intriguing and incredible.

To understand how conscious dynamic flows in the human being, it is necessary to divide it into two components (as indicated at the beginning of the chapter): *«level and content of consciousness. »* The first is closely related, although not totally, to the waking state, that is, being awake or not. The second component, the content of consciousness, represents the sum of all the higher mental, cognitive, and affective functions that allow the knowledge of the external and internal world and the interrelationships between both. It is essential to note that most of our brain processes are non-conscious, and consciousness is the tip of the iceberg of brain activity.

From a thermodynamic point of view, the greater the number of interactions between brain networks, the greater the degree of associated entropy. This phenomenon is due to the more significant number of synapses involved, for example, in the alert state. At the same time, the entropy is lower when fewer combinations of connections occur, which is characteristic of unconscious states or altered alert conditions, for example, when having the eyes closed.

These changes in the degree of entropy depend not only on the mental states but also on the degree of complexity of the neural networks involved. The greater complexity of the connectome of the nervous system, the greater the entropy, and the greater the optimization of information processing, and vice versa *(Guevara-Erra, 2016)*.

Phylogenetically, consciousness is distributed throughout the animal scale, not only in mammals and birds, and even octopuses show a remarkable ability to learn, understand, and remember. The presence of a more complex brain with stronger hierarchical connectivity correlates with more complex cognitive processes and potentially higher levels of consciousness *(Boly, 2013)*. Human cognition is unparalleled in the animal kingdom because it is more than the sum of its parts and depends on the synergy between a unique combination of symbolizing and cooperative traits and prosocial motivation features that nurture many of our most notable cognitive achievements. *(McLean, 2016)*.

However, even insects show evidence of cognition *(Webb, 2012)* and a primitive consciousness. Furthermore, some researchers have suggested they could have subjective experience *(Barron, 2016)*. This assumption is unlikely because subjectivity is a property of perceptual experience and implies a mind that generates a personal identity, representing the "self" from which subjectivity arises. Therefore, it is easy to guess; it requires a complex and evolved nervous system *(Key, 2016).*

For the ancestors of plants, *cyanobacteria* could sense and perceive the ubiquitous distribution of light energy. Therefore, they learned to take advantage of it in a beneficial way, through a system that converts light energy into chemical energy, *«photosynthesis. »* Nevertheless, plants have an osmotically active structure requiring a relatively rigid and large mass that limits their movement. With millions of self-organizing cells and complex control systems, the individual plant allows regional environmental knowledge and exploitation through its biological design. Therefore, plant consciousness is anatomically non-localized and distributed throughout its entire structure in contrast to the more centralized location in the animal brain; however, it includes sensation, perception, planning, learning, and memory.

Plants clearly show a sense of the environment, later integrating it as perception to direct their growth and movement towards the sunlight and their roots in the direction of water and other necessary nutrients. These responses are very fast in some species, such as bamboo and sunflowers. On the other hand, in densely populated areas such as jungles, they compete to achieve the most prominent space of light radiation through various strategies.

Its roots have a conical coverage surrounding their apices called "calyptra," a dynamic structure containing statolith plastids or gravity-sensing plastids that guide root growth downward and to the sides. In addition, when the tips collide or come into contact with objects or obstacles such as stones, it triggers an unusual *"dog-leg"* structure in the distal growing region, placing the tip at an angle, enabling it to slide over an obstacle surface to keep advancing.

Another phenomenon closely related to perception is observed when a phosphate deficiency is in the soil. Signals are transmitted to change the morphology of the root, and when an excess of nitrates is present, the growth process stops. A similar phenomenon occurs when salinity increases and droughts *(Trewavas, 2016).* Thus, plants also show their intelligence by adapting and interacting with their environment, exhibiting the necessary behavioral ability to achieve their purposes successfully *(Calvo, 2020)*.

On the other hand, there is another fantastic discovery. The plants, especially trees, form huge symbiotic communities with fungi, and both kingdoms cooperate for their survival and evolution. Mycorrhizae — fungus *(mycos)* and roots *(rhizos)* of plants— are associations made up of a set of fungal hyphae called *"mycelium."* When in contact with the roots of plants and trees, that can wrap them, forming a large mantle that interconnects underground the roots of the plants, be they of the same or different species, achieving a free flow of nutrients.

Plants provide the mycelium with some carbohydrates that they generate through photosynthesis. At the same time, the latter enzymatically decomposes the soil to facilitate access to nutrients so that the roots of the plants can reach them *(Schwartzberg, 2019)*.

The plant species that form mycorrhizae have different physiology and ecology from those that do not develop this link. Therefore, the mycorrhizal association is considered one of the promoting factors of plant diversity by increasing plants' survival, growth, and reproduction even under environmental stress conditions *(Camargo, 2012)*.

All this complexity is due to the formation of vast networks of mycelia that reach up to 10 km^2 (6.2 sq mi). They also function as an excellent communication media to send various signals, including warnings *(Xie, 2019)*, similar to the central nervous system, thus creating a shared environmental awareness between fungi and plants *(Fig. 3.8)*.

Of course, several researchers reject that other animals, plants, and fungi also have a conscience besides humans. However, subjective human experience is just that, "human" and other living beings possess their conscious knowledge at their "own" level. Although this experience is less extensive due to the lower complexity of their nervous system, we should not pretend that consciousness as such —its meaning and significance— are exclusively related to the nature of humans. As we have already seen, only single-cell protozoa have a sufficient capacity to sense and learn, a primitive way of perceiving their environment. Consciousness and intelligence are undoubtedly at the service of survival.

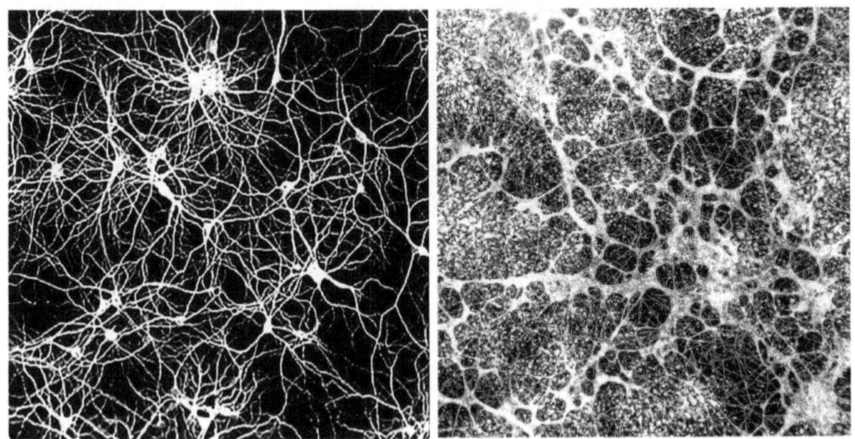

Figure 3.8   Left: Neurons of the human brain.   Right: Mycelium.
Credit: Xie, S. Mycelium Book (2019). Project Proposal.
https://biotechdesign.artscinow.com/node/1556

A recent reductionist trend affirms that the primary function of the animal consciousness supported by the sense organs is to benefit the somatic nervous system that the musculoskeletal system represents to survive. Thus, conscious information allows deciding between fighting or running *(Morsella, 2016)*.

In summary, we have briefly examined consciousness's evolutionary biological bases. The concept of consciousness is elusive as a whole. Locating it from a holistic point of view is very complex, and that is why it requires physical, biological, and psychological analysis. These points of view lead us to know different facets of its essence, some with more solid foundations than others. From a sublime perspective, it has been said that consciousness is how the Universe gives an account of its nature and existence, and perhaps, just perhaps, is the ultimate end of life to exist.

Now, returning to the controversy about whether plants have consciousness or not, it seems to have an affirmative answer by verifying the concepts of quantum consciousness. Primarily, the theory of the *"Orchestrated Objective Reduction"* by Stuart Hameroff and Roger Penrose indicates that consciousness occurs if a system is appropriately organized and capable of developing and maintaining a coherent quantum superposition until a specific goal is reached (what?). This system is present in all plant and animal cells, especially neurons *(Hameroff, 2014)*. Furthermore, with this assertion, we will go to the next section and explain the bases of quantum physics in the generation of consciousness.

Quantum Basis of Consciousness

By including the word "quantum" associated with the subject of consciousness, it seemed that I was taking the risk of being misunderstood. That is in the sense that today many representatives of pseudosciences and bland philosophies have their blogs on Internet domains and use the term with too much familiarity as if they understood what quantum mechanics means, a situation that goes without saying is quite complicated and, in fact, not yet fully understood by even the most brilliant contemporary theoretical physicists. However, I hope you can understand while we develop the theme of why and how this infinitesimal state of matter and energy is probably the origin of consciousness. I want to mention that some parts of this section are very complex, but I will explain them in the simplest and most accessible way.

During the 3rd century BC, Socrates postulated that the brain created consciousness, while Aristotle argued that mental qualities belonged to fundamental reality. In the late Renaissance, Descartes proposed *"Cogito ergo sum"* — "I think; therefore I am."

Unlike other consciousness philosophies, quantum mechanics is much more recent *(Li, 2019)*. It is possible to find scientific bibliography about it for almost a century; however, it was not until the 1990s that the factual bases that explain the phenomenon of consciousness were developed based on theories of quantum physics. In the early 20th century, the original motivation for relating quantum theory to consciousness was essentially philosophical. Free will cannot exist in Laplace's deterministic world, so quantum randomness opened up new possibilities for free will, although it compromises in some way volition.

In the late eighties, some theories related to synaptic bonds with specific quantum mechanisms were published *(Penrose, 1989)*. Then, in January 1990, an article was published in Cambridge: *Quantum Theory and the Brain (Donald, 1990)*, from which arises a significant academic current within the neurosciences to investigate the association between higher brain functions with classical and quantum physics, creating a new science: "Neurophysics" as an extension of Biophysics.

Currently, research is being carried out at institutions such as the Max Planck Institute in Germany, the Neurophotonics Center in Montreal, Canada, in the Universities of Oxford in England and New York and Arizona, in the United States. In 2003, a new and controversial scientific journal called *NeuroQuantology* was edited, which published the scientific advances of the collaboration between Neuroscience and Theoretical Particle Physics.

We will begin by remembering the four fundamental interactions or fundamental forces that govern the Universe, and they manifest themselves from the two primordial members —the Fermions and the Bosons. *Fermions* are matter's primary or elemental components, while *Bosons* are the mediators of the fundamental forces that govern these particles. Each force has its intervention in matter and energy at a specific level. For example, the Strong Nuclear Force controls the smallest magnitudes of matter within the atomic nucleus; the *quarks* and their exchange mediators are the *gluons* holding protons and neutrons together.

The Weak Nuclear Force is located at a higher magnitude (slightly larger than the atomic nucleus) and is responsible for *radioactive decay* through the exchange of $W \pm$ *and Z bosons*. Although it operates on matter and energy on a larger scale, its force field is much smaller than the strong nuclear force. The third interaction is the Electromagnetic Force, which refers to the repulsion and attraction *interaction between electrically charged particles*, which causes a flow of electrons to exist. It is responsible for most everyday phenomena such as light, electricity, and magnetism; it exerts its properties at the atomic and molecular level through its mediators, which are *photons*. Remember that the electromagnetic spectrum ranges from the most extensive radio waves to gamma radiation.

Photons can behave either as particles or waves[41]; this phenomenon is called complementarity or duality. Photons are generally emitted as particles, travel as waves, and reach their final destination as particles (quantum decoherence). Still, their behavior can vary in the presence of an observer. Even a single photon can be in several places simultaneously "this is a preamble to what quantum physics is." Finally, the Gravitational Force is the weakest of the four forces. Still, it is the one that exerts its dominance on the largest objects and through the greatest distances. Its mediating energy particles are hypothetical *gravitons* that have never been detected.

Quantum Physics or quantum mechanics is based on the observation that all forms of energy are a magnitude that is not transmitted continuously but is released in packets or discrete units called "quanta." Initially, it was formulated as a hypothesis by Max Planck (1858-1947) 1918 Nobel Prize in Physics and developed over several years by Bohr, Einstein, Rutherford, Schrödinger, Heisenberg, Pauli, Dirac, Von Neumann, Everett, Born, and others. Most of them disowned (especially Einstein) because of its nature, which goes against human reality and logic; however, mathematically and observationally, they themselves ratified it several times.

Many authoritative opinions refer that no one understood and explained quantum physics better than Richard Feynman (1918-1988). Feynman won the Nobel Prize in Physics in 1965 for developing *"Quantum Electrodynamics,"* which describes how light and matter interact, a theory achieving complete concordance between quantum mechanics and special relativity, explaining the dynamics of quantum fields *(Fig. 3.9)*. This man was a remarkable physicist, trained at the

---

[41] Not only photons but all elementary particles of the Standard Model can also present wave-particle duality.

Massachusetts Institute of Technology and with a Ph.D. in physics from Princeton University.

Figure 3.9   Richard Feynman and his Dodge truck where the Feynman diagrams are drawn. They are graphs that represent the trajectories of the particles in the intermediate phases of the collision process. These are the basis for understanding Quantum Electrodynamics.

Public domain.

His doctoral thesis was directed by John Archibald Wheeler, a renowned physicist from Princeton and Johns Hopkins Universities, whom we have already mentioned when speaking of the anthropic principle. However, in his biographical book *"Genius: The Life and Science of Richard Feynman"* writer James Gleick says that Feynman's mathematical machinery was beyond Wheeler's ability.

He was part of the group of scientists on the Manhattan Project, where he performed an important role. However, he also had fun (to the dismay of the military commanders) opening the safes where the secret documents of the project were saved; he was a jazz and samba percussionist and played bongos in schools and bars, even on the roadsides.

He was appointed a foreign member of the Royal Society of London for the Advancement of Natural Science and named the most intelligent man in the world by Omni magazine in 1979. When his mother found out, she exclaimed, "My God, help us!" Feynman was a controversial and brilliant man of science; he participated in the translation of the Mayan hieroglyphs and created the concept of «Nanotechnology» in 1959 under the motto: *"There is Plenty of Room at the Bottom."*

He wrote more than 20 books and many papers in scientific journals. So, for all this, Richard Feynman was indeed a genius and had a prominent place among the greats of Quantum Physics.

Quantum Mechanics is the branch of Physics offering a fundamental and rational description of nature at extraordinarily small spatial scales. A nature with bizarre reality for the level of perception to which we are accustomed, but its existence of the almost infinitesimally small is how matter and energy behave at their fundamental levels. It is also the physical foundation of everything that exists; however, only some of its manifestations can recognize in our human dimensions. In most cases, the events we observe daily are better explained by the laws of classical physics.

The outstanding components of quantum mechanics include, in addition to the transmission of energy in packets, the already mentioned complementarity or particle-wave duality of the photon, which means that particles and waves are interchangeable, and not only that, but they can also overlap, that is, to exist simultaneously in the two states. These are probability waves, so they are distributed through space in a non-localized way to be everywhere simultaneously. This quantum phenomenon is why Rutherford's model of the atom, similar to a miniature solar system, is wrong since the location of the electrons is not always in a particular orbit like the planets around the sun but which are actually in a cloud of probabilistic waves around the nucleus.

Electrons have an associated wave, and only orbits in which an integer number of "electronic wavelengths" fit are possible. That would explain why some orbits are valid, and others are not. The atoms' stability is why the electrons do not fall into the nucleus (emitting electromagnetic waves) existing in an orbit of minimum energy. Thus, it is substantiated that the atomic model of Bohr and Schrödinger is the correct one.

The wave-particle duality occurs not only in photons but also in electrons that can have a wave nature, as demonstrated by Louis-Víctor de Broglie in 1924, which allows them to be located in many places simultaneously.

Likewise, from quantum chromodynamics, each particle, like quarks, can have many colors simultaneously, but they interact if two particles are generated together. An entanglement phenomenon occurs when one of the particles has the same spin[42] or color as the other particle, it instantly changes to another spin or a different color. The two particles can change at will, for example, from red to green or green to blue. Still, they never keep the same color or spin simultaneously even though they are at enormous distances, and these qualities are called phenomena of non-locality and non-separability.

This entanglement is the essence of quantum coherence and the *Pauli Exclusion Principle* (two particles cannot have the same quantum state). This interaction occurs instantaneously, even faster than the speed of light. It is exclusive for Fermions but not for Bosons since they permanently retain their probabilistic wave property, and individuality does not exist to them.

Still, instead, bosons converge in quantum complex states, and this "convergence" is the foundation that explains the phenomenon of superconductivity when associated with intense magnetic fields are currently the technological base of medical diagnostic equipment called Nuclear Magnetic Resonance and also of high-speed train technology *(Maglev)*. On the other hand, the quantum coherence of fermions is the foundation for developing quantum computing and advanced cryptography.

Another quantum phenomenon includes the Heisenberg uncertainty principle that we already referred to, which explains that the more accurately we can measure the speed of a particle, the more difficult it will be to determine its position in space and vice versa. That is because

---

[42] Spin is the intrinsic angular momentum of a particle or a system of particles.

each particle can have several simultaneous speeds and positions and carry different amounts of energy simultaneously *(Crespo, 2017)*.

Uncertainty in energy raises the possibility that for short times can circumvent the first law of thermodynamics or conservation of matter and energy, and particles can be created, literally out of nothing, which is a property of quantum fields. Finally, the quantum tunneling effect means the quality that particles and waves pass through an energy barrier. It is a consequence of the particle-wave duality. This effect can be observed with radioactivity, radio waves, light, and the tunnel-effect microscope that captures subatomic images by "tunneling" electrons. But this effect under normal conditions with baryonic matter and the dimensions of a human being has a probability of literally zero *(Santaolalla, 2016)*. In *Figure 3.10*, we can see a simplified diagram of quantum fields and phenomena.

Going back to the interactions of the four forces of nature, it turns out that these predominate in objects according to their mass and structural size. We have already talked about the strong and weak nuclear forces and the ultramicroscopic levels at which they work. But for example, in the dimensions of the insect world, what predominates and, by far, is the electromagnetic force over the gravitational one *(Cox, 2013)*. That is why insects walk on walls and ceilings without any problem, and it is also true that a tiny insect, let us say an ant, can drown inside a drop of water on the ground since the drop by the cohesion of its molecules or surface tension, retains its rounded shape (with the ant inside).

On the other hand, Newton's Law of Gravitation is not Universal; it is probably valid only up to tens of nanometers *(Will, 2014)*, but for us humans, this is the predominant force in our daily lives, so within our magnitude and greater intensity in larger masses such as mountains, planets, stars, and galaxies it is the gravitational force that governs with much greater power.

Finally, at the level of clusters, superclusters of galaxies, and the entire Universe, it seems to be that it is dark energy with its repellent nature that governs its evolution and destiny. Some theories suggest that the dark energy will progressively advance over time, manifesting on smaller and smaller scales, even subatomic, to cause the hypothetical *Big Rip* of the entire structure of the Cosmos within several billion years.

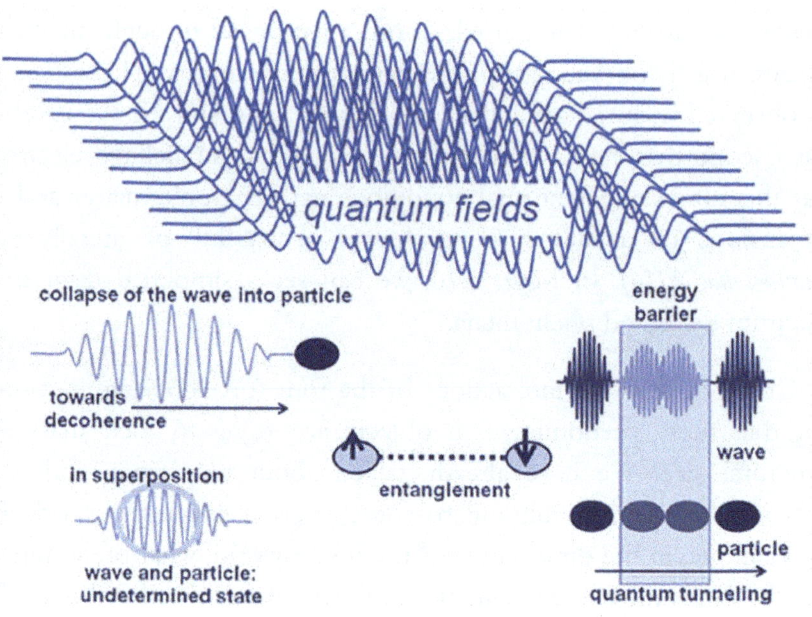

Figure 3.10   Quantum fields and related phenomena.

Modified from Korf, J. (2015) Quantum and Multidimensional Explanations in a Neurobiological Mind Context. *The Neuroscientist* 21 (4): 345-355.

The physical dimensions of human beings are not the best level to understand the reality of quantum physics. The perceptions we experience daily are, in most cases, totally different from quantum ones. However, we can often observe unpredictable or chaotic variables derived not only from classical physics but also from quantum mechanics. From the most basic such as the movements of a double pendulum to meteorological phenomena such as hurricanes, earthquakes, volcanic eruptions, tornadoes, giant rebellious waves, and even much closer to us as is the unpredictability of our existence, where and when we are born, where we live and work, whom we marry with, or when we die.

Again, the genius of Albert Einstein, who seemed to have had no limits, was the first to raise this point in an article published in 1917 with the following question: *What would the classical chaos which hides everywhere in the whole world do to quantum mechanics, the theory that describes the atomic and subatomic worlds? (Gutzwiller, 2008).*

From that question, forty years later, a theory called Quantum Chaos was developed, whose theoretical and mathematical approach is highly complicated, and in fact, no real solution had been found until recently. Some physicists-mathematicians from the University of Lund in Sweden developed a universal system to describe the behavior of chaotic systems at the quantum level. This system has been concretized in an exact mathematical equation that explains the phenomena in quantum chaos at the atomic level *(Riser, 2017).* One way to interpret some realities' counterintuitive and irrational behavior is to understand that under that unpredictable behavior, specific underlying quantum mechanics phenomena exist.

Although not only in irregular systems such as chaotic ones, also each ordered system is made up of smaller subsystems that can be decoupled entirely and at the same time concatenated. Let us remember that quantum physics is the existential foundation of all matter and energy in the Universe *(Gutzwiller, 2008).*

Now, the best way to transfer the phenomena of the quantum world to our reality is through *Statistical Physics*, which is a branch of physics that, using the «Probability Theory, » can deduce the behavior of the macroscopic physical systems based on a statistically significant amount of the elements or particles that integrate them and their mutual interactions.

As we already mentioned, the most disturbing and, at the same time, fascinating aspect of quantum mechanics is that the world does not appear to be quantum physics. When we see something, it seems to have a location, not an overlap of all possible sites. When we travel from one place to another, we take a path, not the sum of all ways. Albert Einstein expressed one of these similar puzzling situations quoted by the physicist and historian of science Abraham Pais: *"I remember that during a walk Einstein suddenly stopped, turned to me and asked if I believed that the Moon exists, only when he looked at it."*

The conventional quantum-physicist answer or «Copenhagen Interpretation» to that question would be: *"Sure, the Moon exists when you are not looking at it, although nobody knows how or why[43], since consciousness creates reality"* (Carroll, 2008).

There is increasing evidence that quantum mechanics is essential to understanding some biological phenomena such as energy transfer in photosynthesis, birds' navigation using the Earth's magnetic field, and the tunnel of electrons and hydrogen in enzymatic biochemical reactions. However, extending quantum mechanics to all functions of biological matter, including cognitive processes, poses a significant challenge. For example, all chemistry, including biochemistry, is based on the creation and destruction of bonds between atoms and, therefore, in quantum interactions, so living systems such as non-living systems depend on quantum states at the level of chemical bonds. The same is for biochemical reactions in the brain, such as neurotransmitters that bind to receptors to send signals through neurons. Furthermore, the quantum states organized in cells are integrated among their

---

[43] George Berkeley (1685-1753) was a distinguished Irish empiricist philosopher and Anglican priest from whose last name the city and the University of Berkeley in California took their name. In his main work, *"Treatise on the principles of human knowledge,"* he mentions that perception is the only thing that we can know; when you talk about an object, you are actually talking about the perception of the object. In connection with a similar question asked about whether a tree exists when no one perceives it, his response was, *"the tree continues to exist when no one apparently perceives it, simply because God is constantly observing it."* The "Copenhagen Interpretation" (because of Niels Bohr's birthplace) of the quantum world developed by Heisenberg and Bohr states that there are questions about the microscopic world that do not make sense. Only questions that affect macroscopic observations will make sense. The ultimate reality remains veiled *(Ramanujan, 2013).*

components; hence the collective conditions of cells can lead to quantum entanglements between them and quantum coherence between tissues and organs, especially in the brain *(Hameroff, 2014)*.

Unlike molecular biology, born in the fifties, quantum mechanics cannot yet explain with the same elegance the processes of life and the mind as it has already described other states of matter and energy so obviously. Unfortunately, biological systems are so large and complex compared to standard physical systems that it is difficult to separate quantum elements from the large number of essentially non-quantum processes that are also present. In relation to this problem, precisely the solution lies in identifying and delimiting which processes quantum decoherence occurs; that means that a system under certain specific conditions stops exhibiting entangled quantum effects and passes to exhibit typically classic physical behaviors. This characteristic happens through interaction with the environment through the fundamental forces of nature or by interaction with the consciousness of an observer.

Quantum neurobiology was coined in the last decade of the 20th century. It refers to the reduced space where quantum physics operates within the nervous system, giving rise to the appearance of higher cognitive functions such as consciousness, memory, internal experiences, and election and decision-making processes *(Penrose, 1989; Jedlika, 2017)*.

Several investigations have demonstrated that photons or quanta of electromagnetic energy can be absorbed and emitted by DNA molecules, especially low-intensity ultraviolet rays. It is also true that living cells perceive infrared electromagnetic waves with a sensitivity peak close to the wavelength of 1000 nm.

Mitochondria, by transferring protons involved in energy production, release near-infrared photons, and in counterpart centrioles, which have been called "the eyes of the cell," are structured to absorb these photons and trigger a signaling cascade. The centriole plays a crucial role in the organization of cellular activities. In addition to being the eye, it is probably also the brain of the cell or the "cell control center," The intracellular microtubules with which it is connected are the mediators between this organelle and autonomic intracellular domains and with other neurons. This organelle detects objects and cells by pulse signals close to infrared, which is a type of proto-consciousness at the cellular level. It occurs when a cell receives signals from its environment, feels them, analyzes them, and reacts appropriately.

All these electromagnetic signals and at least part of this processing involve quantum effects. Therefore, since 1942 a *"Theory of the electromagnetic field of consciousness"* was proposed. The theory shows that the synchronized discharges of neurons correlate with the formation of these fields, giving rise to perception.

The brain's electromagnetic field corresponds to the integral manifestation of all neural information and has dynamics equivalent to the phenomena expected for the integration of consciousness *(McFadden, 2002)*. Recently this thesis has been proven, relating the action potentials in the ionic channels of the neuronal membranes with the "perceptual fields" that are observable as *"qualions,"* which are virtual components of bosons and endogenous electromagnetic fields. These are the theoretical constituents of the essence of consciousness or "qualia" *(Hales, 2014)*. There is also the hypothesis that there may still be undiscovered quantum fields that, unlike known fields, can induce subjectivity when they interact with certain information from the brain *(Schiffer, 2019)*.

Biological systems generally operate under the laws of classical physics; however, it is essential to note that cellular metabolism works under the statistical quantum regime[44]. This is crucial since metabolism is a primary constitutive condition of life. Energy transduction in living systems involves three primary modalities: Photosynthesis (in plants

---

[44] For statistical physics or mechanics, individual quantum events without the mediation of an observer are random, but the behavior of the sets of such events can be determined statistically.

and some bacteria), Ionic Gradients (in mitochondria and chloroplasts), and Oxidative Phosphorylation or glycolysis (in animal cells). If all energy transduction processes in living systems involve quantum mechanisms, then these become an inherent property of living matter. For example, enzymes naturally carry the quantum tunneling effect to catalyze chemical reactions, transferring electrons and nucleons such as hydrogen and deuterium.

On the other hand, the efficiency shown by our brain is surprising. A minuscule 25-watt power consumption produces a colossal information processing capacity close to 10^{20} flops, typical of quantum mechanisms, compared to the power consumption of a supercomputer that consumes around one Megawatt (10^6 watts) with a processing rate of 10^{15} flops *(Sahu, 2014)*.

For many years, experimentation in quantum physics has been known to be exclusively probabilistic. Therefore, anyone who carries it out will not simply be an observer of the phenomena that occur but rather "a participant," and there is evidence that any observer's expectations can change the experiment's final result and that consciousness interacts with random physical processes.

Recent studies confirm that quantum objects seem to have "knowledge" that they are observed and show a change in their behavior according to the experimenter's expectations. This phenomenon is probably the most relevant and extraordinary aspect of quantum physics. This effect is called "collapse of the wave function or elimination of quantum states," which means transforming a quantum object into an object of classical physics [45] *(Tarlaci, 2016)*. It occurs when a wave function initially in a superposition of several inherent

---

[45] This concept is based on Young's experiment of the passage of electrons through a double-slit or slot. Suppose the experimenter observes the electrons passing through one of the slits, knowing that they did indeed pass through that slit using a probing interaction. In that case, the observed pattern will be of a defined path through the slit striking the screen in the form of two bands directly proportional to the grooves, as expected for the behavior of the particles. On the other hand, in the absence of direct observation, as they pass through one slot or another, the screen shows an interference pattern that indicates the wave nature of the electrons. In practical terms, "collapse" happens when an observer wants to know which path an electron has taken.

states reduces to a single characteristic state due to interaction with the external world.

All of this entails essential consequences for science and life:

1st   physical reality can be created by consciousness,

2nd   quantum objects are subordinate to the will or intentionality of any conscious observer, and

3rd   consciousness for quantum physics goes beyond the electro-chemical processes of the brain and, therefore, transcends them.

This results in consciousness as part of the nature of the Universe, having the ability to create realities. These effects during experimentation are even more evident in people who do meditation and therefore have a higher ability to focus their attention *(Radin, 2012; 2013)*.

Niels Bohr affirmed that we could not attribute properties to quantum objects unless these have been measured; nevertheless, a quantum particle that is everywhere only appears in a specific place until an observer determines it. This will be presented in the form that the observer wishes, called in quantum physics «problem of measurement or the measurement. » It is impossible to know the properties of some quantum particles except when it is present according to the observer's intentions. Bohr wondered, *« how do we know that a quantum object exists without measurement or observation? »* —The answer is that nobody knows. Still, we can deduce that human beings should be considered the ultimate measuring instrument. If so, what is the nature of human beings that gives us this exclusive quality? A little reflection will make us think that human consciousness is unique and different or at least more complex.

However, it would be reasonable to assume that possible alien consciousnesses equal to or even superior to ours, in a unified way with humanity, could be the creators of the reality of the Universe. However, even being more acute, why locate only the higher consciousnesses if it has been verified that animals, plants, and even every quantum particle could have "consciousness." Then we can conclude that everything that exists constitutes Universal Consciousness or, seen in another way, the very cause and essence of the Universe is consciousness. That is called *Panpsychism*, which we will review objectively later *(Mocombe, 2019)*.

Some opinions refer that quantum states, when materializing (quantum decoherence), do not simply become electrons, protons, or neutrons but will be transformed into desires, intentions, and objectives within the mind of each human being through the collapse of the wave function. But only, if this collapse occurs first within the human mind, it can also happen outside, and that goal or desire will materialize *(Ramos, 2016)*. This could be the explanation of the famous reflection by Emerson and later by Paulo Coelho quoted in his work "The Alchemist" *«when you really want something, all the universe conspires to help you achieve It. »*

It is necessary to interpret these concepts judiciously since, in themselves, they are part of what has been called "quantum mysticism," which are pseudoscientific metaphysical beliefs that incorporate Oriental and "New Age" philosophies, supported by characters such as Deepak Chopra, who within his statements and misusing quantum mechanics declares in his work *"Quantum Healing"* that the mind causes aging. In 1998 Chopra was awarded the Ig Nobel Parody Prize in the Physics category for "his unique interpretation of quantum physics and its application to life, freedom, and the pursuit of economic happiness." In this regard, Barbara Ehrenreich, social critic and investigative journalist mention: *'I love that at the expense of quantum physics, for some reason, it has become an excuse to make fun of all science. Seeing that nothing is real, nothing is true, or what you think, this is how the world is. So, if you think positively, you remake the world positively according to this pseudoscientific explanation".*

These types of comments, as well as serious and objective criticisms, force us to be cautious in how we analyze proposals and evidence. Recent research, for example, has not been able to establish an association between prayers and intentions, with the healing of sick people, except on some occasions when the patient is aware that they are praying for his health, and probably depending on expectations and motivations of the participants in the prayer groups *(Schlitz, 2012; Radin, 2015)*. That could confirm that if the collapse of the wave function does not occur in the patient's mind influenced by the prayers of the prayer groups, then the collapse will not happen outside their mind, and there will be no improvement or healing.

This particular mystical question is transcendental because the human being, traditionally in all cultures, when confronted with serious illnesses, his immediate response, in addition to seeking a cure, is to raise prayers. However, it is essential to point out that a healthy scientific skepticism is vital to restrict the physical from the metaphysical features to separate the hypothetical from the verifiable and the beliefs and atavisms of the facts, making an honest effort to arrive at the truth, even if it goes against our most fervent desires and aspirations.

Quantum physics, really at the current level of its knowledge, cannot universally explain the phenomenon of consciousness, and therefore, most of the testable concepts are, for now, grounds of classical physics and biology, which indicate a central point that the level of consciousness is directly related to the complexity of the nervous system. However, the advance in understanding quantum mechanics applied to biological systems in the future could explain deeper how this phenomenon originates. Finally, all matter and energy owe their existence to the waves and particles that constitute them.

Wave functions are physically objective so that the micro-physical systems can exist simultaneously at the quantum level in more than one state and place. But once the threshold of the collapse of the wave function is exceeded, the decoherence appears with the consequent formation of macroscopic systems endowed with mass that, therefore, cannot exist in more than one place at the same time. Therefore, all macroscopic systems, whatever their nature, are determined and unified by the same underlying mechanism of the objective collapse of the wave function that gives rise to them.

In Hugh Everett's proposal of the Multiverse proposal of Infinite Worlds and Minds, there is a branched, divisive wave function. Reality is continually separated into infinite and different consciousnesses, which have no knowledge of each other within the Multiverse.

Now, to understand that quantum phenomena are the basis of consciousness, we must look for them in their constituent elements: first is *sensation* followed by *perception,* and finally, *learning* and *understanding;* David Chalmers would say, «to start with the easy problems of consciousness» *(Chalmers, 1995).* Remember that sensation occurs in the sense organs; specifically, in two of them, their quantum foundations have been demonstrated: sight and smell.

Sight, without doubt, is for human beings, the primary sensory organ through which the most significant amount of input enters our brain. The eyes are transducers that convert electromagnetic energy (photons) into action potentials understandable to the brain. When the light is reflected on objects, it is focused on the retina, where the pigments of the photoreceptor cells are activated to give the electrical signals that will produce the visual sensation.

Observations indicate no difference between a photon impacting the retina and the quantum photoelectric effect; that is because the brain can perceive the micro-stimuli of only 2 or 3 isolated photons[46]. These photons reach the retina in the form of bundles or "quanta" of light, which activate the rods and cones in our eyes, and in a way not well understood, the brain interprets the illuminated objects *(Fig. 3.3)* and transforms them into intelligible images (perception).

The color with which we see an object depends on the spectral composition of the reflected light *(Fig. 3.11)* and represents the image

---

[46] The photoelectric effect refers to the emission of electrons when light is reflected off a material. Electrons emitted in this way are called photoelectric electrons, and reciprocally, photons are emitted when an electron jumps into a lower energy orbit in an atom *(Sliney, 2016).* Albert Einstein gave the theoretical explanation for this effect, which earned him the Nobel Prize in Physics in 1921.

that gives our eyes through a system of trichromatic vision that we possess.

The vision of human beings is trichromatic because the 6 million cones that the retina has are of three types *(Fig. 3.12)*; those that capture long wavelengths "L" (red light), those that perceive medium wavelengths "M" (green light), and finally those that are sensitive to short wavelengths "S" (blue light). The wavelengths capable of producing visual sensation range from 360 to 830 nm in young adults and 400 to 760 nm in older adults *(Sliney, 2016)*.

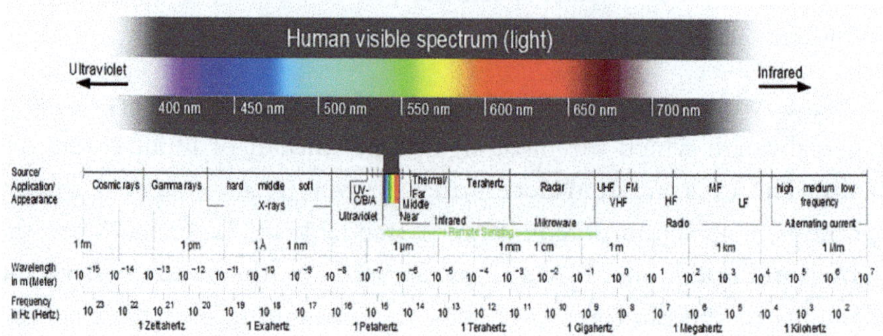

Figure 3.11   Electromagnetic spectrum.
The small visible area of the human eye is amplified.

Credit: Horst Frank. Public domain.

Recent investigations show that humans are able to detect infrared light between 950 and 1150 nm and perceive it as visible light. In fact, the perception is reported as a pale green light; this is because the amount of energy emitted towards the retina by two photons of that wavelength is equal to a single photon of about 550 nm, and that wavelength is precisely that of the green color.

These observations are consistent with rhodopsin's energetic quantum mechanical model of two-photon activation *(Palczewska, 2014)[47]*. If two photons emitting simultaneously are entangled, they must be considered as a single quantum state, a single wave function.

---

[47] Rhodopsin is a retinal rod photoreceptor protein in low light conditions.

The human retina's resolution with its 100 million photoreceptor rods is inconceivable. It is equivalent to a 576-megapixel camera with an ISO sensitivity of 800 *(Clark, 2016),* and the cones have the ability to distinguish more than 2.2 million shades of color *(Masaoka, 2013).* Therefore, the perception of visual sensation in humans is performed mainly in the primary visual cortex located at the posterior pole of the occipital lobes of the brain.

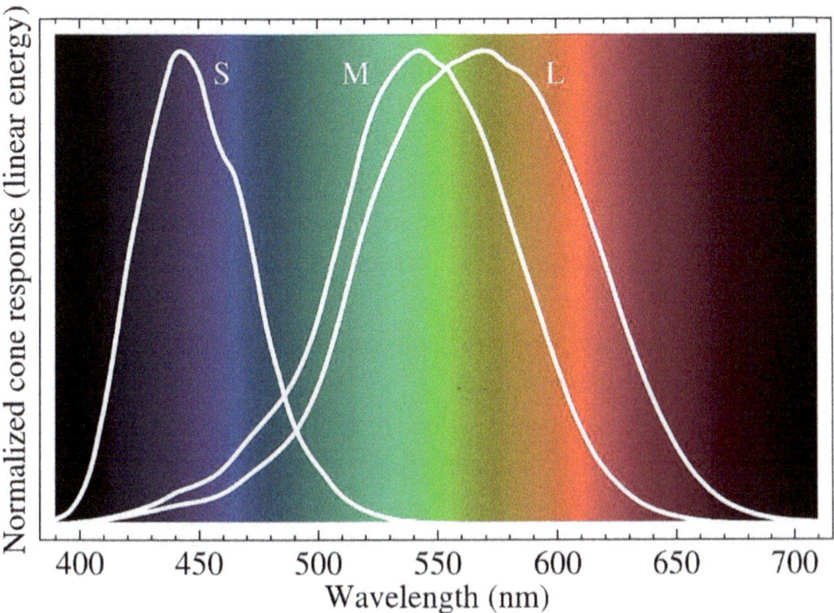

Figure 3.12   *Normalized response spectra of human cones: S, M, L.*
(Trichromatic vision)
Credit: Ben RG. Public Domain

Most animals of various phyla, for millions of years millions can also see ultraviolet light, which means they have a *tetrachromatic vision;* among them are fish, insects, amphibians, birds, reptiles, and mammals, including dogs and cats, but the great apes, including humans, have lost it. Ultraviolet vision fulfills a specific function for each species for survival purposes; for example, bees are guided to the nectar of flowers, and reindeer can see polar bears despite being in a snowy environment; their range of UV vision goes from 350 to 320 nm.

Animals like humans who cannot see UV light have shown higher resolution vision. It is unknown how it happens, but this would mean suffering an adaptive evolutionary loss in order to have a visual capacity in greater detail. On the other hand, good night vision is essential for the survival of many animals, depending on the presence of a reflective layer located at the back of the eye called *"tapetum lucidum."* This bright mat is situated behind and in the middle of the retina, it allows to increase by reflection the amount of light directed towards it, thus improving its sensitivity to low light. Dolphins, whales, cats, canids, rodents, cattle, and others are examples of animals that have this visual advantage.

Some other species not only have a greater visual capacity, but most migrant animals (of all classes and orders, including insects) have in their eyes special receptors that capture the geomagnetic field as a biological compass that allows them to orient themselves in time and space for their journeys. Cryptochromes in the retina of birds appear to be the primary sensory molecules involved in the orientation mechanism, sometimes present only in the right eye and in nocturnal migration species, usually in both *(Engels, 2012)*. Capturing the Earth's Magnetic Field through cryptochromes in the eyes of certain species is feasible because they can quantumly entangle several electrons whose spins act as cubits. This means that for millions of years, nature has been working by quantum computing *(Gauger, 2011)*, a situation that, for a man, still represents serious difficulties, and the design and construction of computers of this type are hardly in their infancy.

We conclude then that with respect sense of sight, there are at least three apparent quantum phenomena:

1. *the photoelectric effect.*
2. *the energetic model of the activation of two photons, and*
3. *the entanglement of electron spins that act as cubits.*

These mechanisms present in the sense of sight an excellent start to understanding the integration of some of the underlying quantum components of consciousness.

In the case of the sense of smell, the biophysicist Luca Turin in 1996 proposed a theory to include quantum mechanics in olfactory sensation and perception.

He postulated that the olfactory receptor contains electron donor and acceptor units that are energetically separated by a space that coincides with the vibration quanta of a specific odorant, and this is called a *phonon*. When an odorant binds to its receptor, a quantum electron tunneling event occurs, which in turn initiates transmission access to the brain. Olfactory receptors are closely bound to zinc receptors, and this element can be seen as the donor source of the electrons needed for quantum tunneling. That is why we can distinguish an impressively more significant number of different smells or perfumes with a limited number of odor receptors that are approximately 347. It also explains why molecules with a minimum change in their chemical structure have a distinct odor *(Brookes, 2010; Trixler, 2013)*.

The knowledge for many years was that the olfactory capacity of the human being amounted to $\pm$ 10,000 aromas. In March 2014, an experimental study was published in the journal *Science* entitled "Humans Can Discriminate More than 1 Trillion Olfactory Stimuli" (1 x 10^{12}) *(Bushdid, 2014)*. This huge number is only possible if quantum phenomena are included in the olfactory process. So, with this information, we can only remotely imagine the number of scents that, for example, a bloodhound can pick up.

Although the quantum phenomena associated with the sense of taste have not yet been described, it is very likely that they will meet in the next few years since smell and taste, although with slight differences, are anatomically and functionally linked. Both sensory stimulations are done by chemical elements (odorants and flavorings). Without smell, the flavor is lost, and perception of food is limited to basic tastes. This interaction between smell and taste that gives rise to flavors occurs together within higher nervous structures such as the anterior insula and the frontal orbital cortex. Diseases such as Parkinson, Alzheimer, and some types of epilepsy, as well as anesthetics such as propofol, ketamine, and lidocaine, and certain respiratory viral infections, produce loss or alteration of both senses simultaneously *(Ribas, 2012; Elterman, 2014; Patel, 2020)*. Human beings have the sensory capacity for six primary flavors: sweet, acid, salty, bitter, umami, and the recently described oily taste *(Running, 2015; Costanzo & Kindleysides, 2017)*; when combining them, we can calculate that we perceive around some 50,000 different flavors in such a way that the olfactory component associated with taste perception is probably carried out at least in part by the same quantum phenomenon of electron tunneling.

As there is no solid evidence for quantum phenomena for the sense of taste, neither for the moment are there for hearing and touch. However, concerning hearing, it can also be verified as an underlying quantum phenomenon in the future through research to observe the strange behaviors of many animals before earthquakes and tsunamis. Although it could be explained due to the perception of low frequency (below 20 Hz) seismic sound waves that are inaudible for humans, as they are for many animals, it does not explain the totality of the observed responses. Attempts to use the altered behavior of both wild and domestic animals to predict earthquakes have a long history. Despite the many occasions on which this unusual behavior has been observed, not a single piece of information has withstood careful scrutiny and achieved a widespread successful prediction, and yes, instead, it is the rule that in all seismic events, anecdotal reports appear in up to seventy species of animals of all orders, especially dogs, cats, rodents, chickens, cows, horses, and even insects.

We must carefully analyze the facts to distinguish, for example: in 2004, the escape behavior of the coasts of all elephants, orangutans, and other animals in Thailand and Sri Lanka due to the arrival of tsunamis in 2004 (with a time of up to 38 minutes before the first wave appeared on the beaches), from the manifestations of escape of millions of ants, leaving their ant hills an hour before an earthquake in Russia, or toads, leaving their hatchery and migrating to higher lands in L'Aquila,

Italy, days before an earthquake in 2009 and returning until the series of aftershocks ended *(Grant, 2011)*.

In the case of toads in Italy, it was shown that showed the chemistry of water in areas around the epicenter (including the hatchery) was modified days before because when the tension between the tectonic plates increases some time before an earthquake, ions such as oxygen radicals are released from the subsoil, which oxidizes the water and causes a part to transform into hydrogen peroxide (hydrogen peroxide) which was detected by these amphibians.

In tsunami incidents such as the one in Indonesia on December 26, 2004, when the final movement of the tectonic plates triggered the earthquake, infrasonic sounds of between 1 to 10 Hz are created and traveled through the air at an average speed of 1,250 kms/hr. (775 miles per hour), while the waves in this particular case traveled at ~ 700 kms/hr (435 miles per hour), which allowed that 1000 kms (620 miles) away as it was in Phuket, Thailand, and Sri Lanka, the animals recognized the warning with more than half an hour in advance.

Based on these data, the correlation between the physical phenomena of tsunamis and the behavior of animals is understandable; however, explaining how animals can predict earthquakes is a different topic. Seismic waves on land travel through the subsoil between 7,200 to 18,000 km/hr. (4,500 to 11,000 miles/hr.), several times faster than sound, so there must be another physical phenomenon that allows anticipation *(Garstang, 2017)*.

Studies and strategies to predict earthquakes are more than 50 years old but have intensified since the terrible earthquake in Japan on March 11, 2011, which had a magnitude of 9.0 on the Seismological Moment Magnitude Scale[48]. In an earthquake, the three destructive variables are magnitude, time, and distance from the epicenter. The latter is the one

---

[48] The Seismological Moment Magnitude Scale is a measure that officially replaces the Richter scale since the beginning of the millennium due to its greater accuracy, especially in earthquakes of greater intensity.

that determines with greater intensity the rare behavior observed in animals, followed by the first variable.

It has been proven that locally —before, during, and after— earthquakes, the formation of large electromagnetic fields causes a current of electrons that, in turn, produce radio and sound waves. The primary electromagnetic emissions (which travel at the speed of light) are Ultra Low Frequency (≤ 1 Hz), originating initially from the earth's crust and later from the ionosphere; this phenomenon is manifested graphically with two peaks, the first one occurs 1 to 2 weeks before the event, followed by 4 to 7 days of quiescence, and then the second peak occurs with greater intensity one to two previous days, progressing to an abrupt increase just a few hours before the event. These observations coincide with the unusual behavior of the animals 7 to 10 days before the earthquakes, with a notable increase in stress in the hours before the phenomenon. The final electromagnetic signal is the strongest and represents the imminence of the quake.

The distances from the epicenter where these changes in animal behavior occur on the Richter scale are: for ~7-degree earthquakes, approximately 100 km (62 mi) in all directions, and 70 to 80 km (37 - 44 mi) for ~6-degree earthquakes; in those of greater magnitude, although there are no accurate records, they can be captured at greater distances, and there is a proportionality. The results of several investigations indicate that the perception of these electromagnetic environmental changes is noticed first by the smallest animals, followed by the largest ones; hence insects, worms, and rodents are the first to feel the imminence of earthquakes *(Hayakawa, 2013)*. Now, these observations lead to how animals can perceive these environmental changes. Before answering, let us see how these strange electromagnetic emissions arise in detail.

The ionosphere is the part of the atmosphere located between 60 to 1,000 kms (37 to 620 miles) above sea level. In this atmosphere area, the gases that compose it, although still linked to the Earth's gravitational field, the space between its components is so large that it can be considered, to some extent, that it is a system of particles without collisions. That gives rise to the existence of ions, mainly oxygen and free electrons. Therefore, the ionosphere acts as a medium that allows the transformation of electron currents into electromagnetic waves within the spectrum, primarily radio waves. However, it also happens frequently within the visible light spectrum from the solar wind that causes the appearance of the northern and southern lights in certain latitudes.

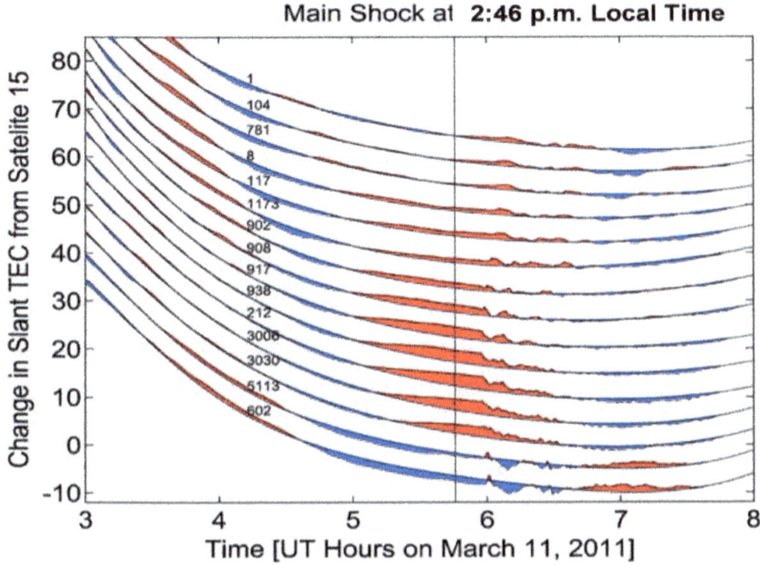

Figure 3.13 *Time sequences of the Slant Total Electron Content in 15 Earth Stations. The areas highlighted in red correspond to the increments.*

Modified from Kelley, M. (2017). Ionospheric total electron content variations before major earthquakes due to electric fields created by tectonic stresses. *J Geophys Res Space Phys*, 122, 6689-95.

For several years it has been observed that the total electron content in the ionosphere from the Earth's crust has increased before and during earthquakes *(Kelley, 2017)*. Specifically, the increase in electrons occurs in those with an inclination pattern alignment but not those of a vertical pattern. Currently, with a large number of ground and satellite stations, exact measurements can be made; for example, in the 2011

Japan earthquake, fluctuations were recorded with increases of up to 10% that began 40-50 minutes before the earthquake *(Fig. 3.13)*.

The increase in the total electron content of inclined patterns *(Slant Total Electron Content – STEC)* distributed in a wide geographic area is caused by the movements of the earth's crust before, during, and after the earthquake. These signals are transmitted at almost the speed of light and are guided by the Earth's magnetic field into the ionosphere.

When fluctuations in the electron content are caused, these electrons are again reflected towards the surface of the Earth in the form of electromagnetic waves called "electrophonic" with a wide range of frequencies. These are the "non-acoustic waves" to which many animals respond[49], what explains why certain species can predict quakes.

The probable quantum phenomenon associated with hearing would be the one in which quantum states are transformed into classical physical states within the inner ear, the auditory nerves, or the brain. This situation occurs through converting electromagnetic waves into action potentials recognizable as hearing by the brain without being produced by acoustic or sound waves. There are many electromagnetic spectrum frequencies through which this phenomenon can occur. It ranges from the ultra-low frequency (≤ 1 Hz) —most related to the alteration in the behavior of animals before earthquakes— up to the range of the ultra-high frequencies located between 300 up 3,000 MHz which is the microwave zone and through which it has been shown that under certain circumstances, they are perceived as sounds such as clicks

---

[49] The wave-photon duality as members of the electromagnetic field propagates through space, carrying radiant energy from radio waves to gamma waves. All electromagnetic waves travel at the speed of light, unlike sound or acoustic waves, which are longitudinal mechanical waves that act by compression in a medium (solid, liquid, or gaseous), which is why they travel at different speeds. Therefore, sound waves cannot travel through a vacuum, unlike electromagnetic ones. Electromagnetic waves are components of quantum physics, while acoustic waves are components of classical physics.

or buzzes even in people with profound deafness *(Frey effect)*, or even as smells in some experimental subjects.

The soft tissues in the head can absorb the pulsation mechanisms of these microwaves. As a result, they launch an acoustic pressure wave that travels by bone conduction to the inner ear, where it activates the cochlear receptors through the same process involved for normal auditory hearing *(Yitzhak, 2014)*. it can also occur as a result of thermal expansion around the cochlea. Hypothetically, it could also occur by direct stimulation by converting the electromagnetic waves (no sound) —within the inner ear, in the auditory nerves, or the brain— into a current of electrons, causing the sensation and subsequently unleashing the action potentials necessary for auditory perception which is nothing more than quantum decoherence or collapse of the wave function. However, so far, there is no evidence of this phenomenon.

Electrophonic hearing is currently used to design and manufacture cochlear implants that achieve auditory sensation and perception in cases of profound deafness, mainly congenital. In addition, electromagnetic and quantum fields have recently been used for auditory perception *(Zamzuri, 2020)*.

As we have insisted, the macroscopic demonstration of quantum events is very complicated and requires elaborate experiments, even though they are almost always present. However, it is not impossible; for example, we are witnessing *Bose-Einstein boson condensates* manifest through magnetic levitation; however, identifying these or other phenomena in biological systems is much more difficult due to their complexity. For this reason, it is truly unique that molecules, structures, and organelles have been located in neurons that potentially house the necessary quantum mechanisms that give rise to cognition and consciousness.

The key to both is the microtubule cytoskeleton in neurons involved in electrical signaling. The microtubules form a Bose-Einstein condensate through which tubulin molecules oscillate inside each cell in conjunction with other cells, forming a single quantum entity, giving rise to the phenomenon of consciousness.

Before continue analyzing these complex concepts, we will review the essential evolutionary points of consciousness.

-First stage: *Sensation* —it means feeling. It is the reception of stimuli or information through the sensory organs.

-Second stage: *Perception* —defining what is felt or interpreting sensations from the environment, and

-Third stage: *Cognition* —is using knowledge to understand what it is through acquiring, storing, and using knowledge. In the human being, cognition includes higher intellectual processes, such as thought, memory, attention, and the complex processes of perception for obtaining knowledge. On the other hand, metacognition is the knowledge of our cognitive processes, the knowledge that each one has about their learning system or style *(Yankovic, 2014)*.

From the quantum point of view, a model has recently emerged that makes it possible to explain the mechanism of cognition. It includes an information storage process based on biochemical phosphorus compounds used during the quantum process of the excitatory neurotransmission of glutamate[50]. The model is based on a technology that is familiar to us, Nuclear Magnetic Resonance Imaging. The images with this technology owe their production to the alignment of atomic nuclei altering their spins (α and β) due to the powerful electromagnetic forces of the equipment.

Of the various atomic nuclei, it is especially the hydrogen in water that is most altered. Water is present in all cells and tissues, so using this technology to obtain images of optimum quality is possible. Nevertheless, another element, phosphorus, allows to create images by altering the spin of its nucleus and is used as an element for nuclear

---

[50] Glutamate is an amino acid, the most important excitatory neurotransmitter in the human cerebral cortex. The other excitatory amino acid is aspartate, its precursor. Both are directly related to the molecular basis of memory.

magnetic resonance spectroscopy, a chemical technique to obtain structural information on molecules[51].

Phosphorous is present in energy production reactions such as ATP and AMP (adenosine triphosphate and monophosphate). Specifically for the quantum cognition process, it has been demonstrated that two types of phosphorus molecules are involved: *phosphate and pyrophosphate.* The action of the pyrophosphatase enzyme on pyrophosphate produces two phosphate molecules that are quantum entangled. The entangled phosphates transported in different neurons can participate in simultaneous biochemical reactions to release neurotransmitters such as glutamate. It results in quantum entangled "non-local" postsynaptic firing among multiple neurons.

On the other hand, for an information storage process or quantum memory to exist, the union of calcium with pyrophosphate is required to form the *Posner molecule* $(Ca_9(PO_4)_6.)$, This molecule is a precursor of hydroxyapatite essential for the mineralization of bones and has the capacity to function in neurons as a cubit for quantum memory. It can also be quantum entangled *(Fisher, 2015; Player, 2018).* The two molecules simultaneously release calcium into the cytoplasm of different presynaptic neurons, which triggers glutamate release. This neural quantum process can provide a model of biological architecture to achieve the development of quantum computing in the same way the knowledge of the network of neural circuits has contributed to the development of artificial intelligence *(Weingarten, 2016).*

In the late eighties, Sir Roger Penrose (2020 Nobel Prize winner in Physics), in his book *The Emperor's New Mind,* adds the quantum effect on human cognition for the first time. He considered that consciousness can affect quantum mechanics and that quantum mechanics itself could be included in consciousness. He also suggested

---

[51] Radio waves have adequate energy to make atomic nuclei come into resonance, subjected to a magnetic field. This technique is called nuclear magnetic resonance (NMR) spectroscopy. The most widely used radio frequency length is between 1 to 5 meters.

that the "Objective-collapse," the collapse with the quantum superposition, is a real physical process, and consciousness is a product of the quantum structure of space-time inextricably related to the Universe. It is the theory that describes the relationship between consciousness and the universe. Later, together with Stuart Hameroff, they refined the quantum consciousness model and called it *"Orch-OR Theory"* (Orchestrated Objective Reduction), which has been investigated since then *(Hameroff, 1998; Penrose, 2011)*.

What does Orchestrated Objective Reduction mean? Reduction refers to the collapse of the wave function, that is, to the passage of an "object" in a quantum state to a state of classical physics and that is orchestrated, harmonized, or computed by intracellular elements, specifically in the cytoskeleton through the complex network of protein filaments (α, β, γ tubulins) that make up microtubules and other proteins associated with these microtubules, such as actin.

The interconnection of the microtubules is carried out through more complex structures —the centrioles. Two centrioles arranged at right angles constitute the centrosome *(Fig. 3.14)*. Microtubules must likely share both states (classical and quantum). For the classical states, the microtubules would express the information in bits, while when they are in a quantum superposition state, they will do so as cubits.

Both classical and quantum computing work in parallel within the brain. The basis of *cubits* is oscillating dipoles that form overlapping resonance rings in helicoidal pathways along the microtubules. Microtubule-associated proteins influence or orchestrate the reduction of the *cubit to a bit state*, modifying the space-time separation of their superimposed states. When these tubulin assemblages are in a coherent quantum state, they can reach a threshold, and from this wave function collapse to a classical physical state is triggered from this threshold, and it is precisely that Penrose has called "Objective Reduction." This self-organized or orchestrated collapse with classical states is proposed to give rise to the «conscious event» and the «real now. » The stream of consciousness would be conceivable as a discrete sequence of quantum "self-collapses" coupled in different real-life situations *(Parrilla, 2016)*. This point is essential because the microtubules would be the precise place where elimination of the wave function takes place for the generation not only of atomic particles but of consciousness itself through information packets that are initially *cubits* and later materialize in *bits* when the collapse occurs. Both information packages act synchronized to generate perception, understanding, and cognition, *conscious experience*. Consequently, we would have to ask ourselves if the objective reduction is the generator of reality or only the way in which we perceive it.

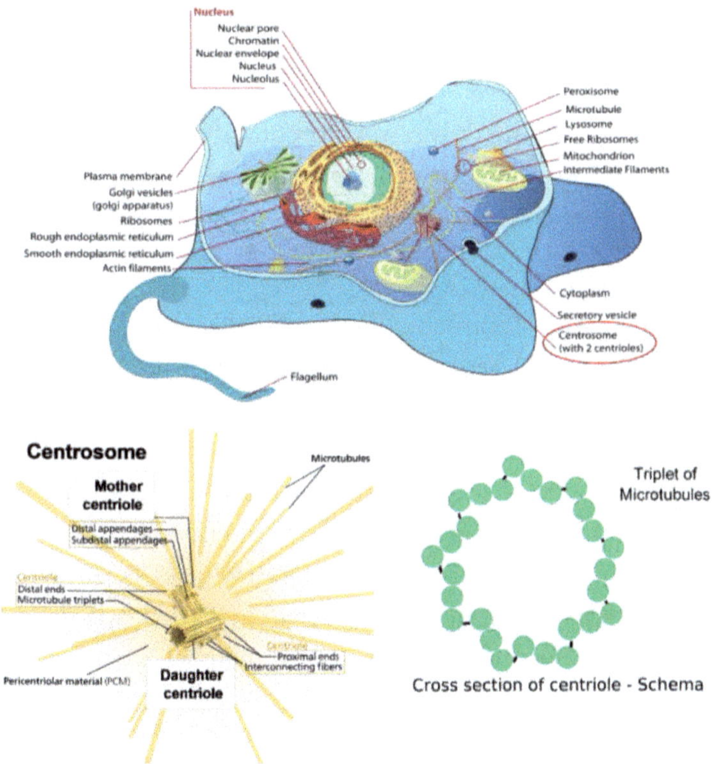

Figure 3.14 Top: Image of an animal cell and its organelles. Bottom-left: Centrosome. Bottom-right: The nine triplets of microtubules of a Centriole.

Modified from LadyofHats – Public Domain. Kelvinsong, Twooars. CC BY-SA 3.0

Several authors point out that it is difficult to verify the quantum elements in the generation of consciousness in the brain since it is a warm, humid, and noisy place, opposite to how quantum phenomena are experienced in the laboratory, which is a cold, dry and quiet space *(McKemmish, 2009)*. However, we have reviewed that quantum phenomena are confirmed in sight and smell and have an environment identical to the brain.

On the other hand, there is the concept that consciousness "emerges" simply as a kind of «epiphenomenon» of brain activity; therefore, it does not matter to specify the exact type of physics that explains the relevant processes.

Therefore, a conciliatory position is that of Computational Functionalism, according to which —it is just the cerebral computational activity that gives rise to conscious experience through the information provided through overlapping *"bits* and *cubits,"* and —*the mind*— consequently, it is the software of the brain *(Piccinini, 2010; Hameroff, 2020)*. [52]

In January 2014, Anirban Bandyopadhyay from the National Institute for Materials Science in Japan verified quantum vibrations in neuronal microtubules, which apparently confirms the hypothesis of orchestrated target collapse theory *(Ghosh, 2014)*. Another study that supports the quantum theory of consciousness was published in 2015 and represented a model that shows that coherent quantum oscillations are the main foundation in the brain's cognitive process *(Korf, 2015)*. In fact, this hypothesis has been confirmed in plants whose proteins actively participate in photosynthesis. The matrices of chromophores

---

[52] An epiphenomenon is a secondary phenomenon derived from another primary or determining phenomenon. It is a by-product of other processes that do not cause anything by itself *(Baumeister, 2018)*. For example, materialists regard thought and consciousness as epiphenomena of brain activity. The phenomena are all the facts involved with our senses. Phenomenal consciousness is a wide field of experiences simultaneously present in a person's same subjective stream of consciousness. A typical example would be the experience of emotional ambivalence that we have in various situations or elections.

(set of atoms of a molecule responsible for its color) create the conditions for quantum resonance transfers to propagate through all chromophores simultaneously.

Within microtubules in neurons, tryptophan aromatic rings can support excitons (quasiparticles) and quantum resonance transfers in a viable way. These types of quantum resonance transfers propagate all along microtubules in multiple neurons simultaneously. Indirect evidence for this mechanism includes anesthetics that act on the quantum channels of microtubules and block these mechanisms, reversibly erasing consciousness *(Craddock, 2017)*.

Recently, the fatty acids of neuronal membranes and serotonin have also been involved as elements participating in the quantum vibration of microtubules that are generators of consciousness. Biochemical peroxidation processes of the fatty acids of cell membranes originate intracellular photons, which are then captured by serotonin and tryptophan[53]. Serotonin can act as a chemical regulator of peroxidation in cell membranes, as it has antioxidant properties.

On the other hand, as we have already seen, the precursor in the biosynthesis of serotonin, tryptophan, has quantum mechanisms inside the central nervous system similar to those carried out in plants. In plants, tryptophan converts solar electromagnetic energy into biochemical energy (ATP and GTP). It provides them with the quantum mechanisms necessary to carry out the specific movements required to head towards the light source, which is undoubtedly a primary type of consciousness. The tryptophan in the microtubules of neurons can capture bio-photons and make coherent quantum resonances through its aromatic rings (benzene and indole). The dynamics of aromatic excitation in microtubules can support possible mechanisms for quantum computing according to specific algorithms

---

[53] Bio-photons such as those produced by mitochondria have a wavelength close to infrared. These will be captured by the centrioles and tubulin dimers in the microtubules forming the hexagonal grids. There tryptophan molecules are inserted.

from which the phenomenon of consciousness could originate *(Tonello, 2015)*.

In the same line of research, cell-to-cell communication through biophotons has been demonstrated in various plants, bacteria, and specific animal cells. Through experimental evidence, it has been suggested that biophotons may play a potential role in the transmission and processing of the neuronal signal, contributing to the comprehension of the higher functions of the nervous system and that the decoding depends on the quantum brain mechanisms. Glutamate induces biophotonic activity in neural circuits related to memory mechanisms. Several authors also suggest that these biophoton patterns may be fundamental for processing neural information *(Tang, 2014)*.

As expected, some researchers question quantum theories of consciousness. The general conceptualization still has gaps in its structure, so it must be consolidated before scientists universally accept them.

In summary, we have mentioned that the energy transduction processes of plants and animals have quantum processes inserted. We also speak of the centriole as a control unit of each neuron and the microtubules as their transmitters for both quantum and physical-classical messages. Likewise, we said that consciousness transcends the electrochemical mechanisms through which the brain functions and communicates. However, the big question that remains unanswered with precision is: *how do quantum mechanisms through biochemical and biophysical systems give rise to consciousness?*

Continuing the topic, we will analyze the duality constituted by consciousness and reality. Although apparently, both are the same phenomenon, it is not, and there is the dilemma of what is the cause and which is the effect. The theoretical elements that support this question are closely related to quantum physics, and these are the Participatory Anthropic Principle and the Biocentric Universe Theory or Biocentrism, whose essential part implies that the observer creates reality.

Consciousness and Biocentrism

To understand consciousness from the perspective of Biocentrism, whose fundament is the Anthropic Principle, we must know some of the ideas expressed in the works published[54] in 2010 and 2017 by Dr. Robert Lanza (doctor and geneticist at the University of Pennsylvania) and Robert Berman (astronomer, director of the Overlook Observatory in New York). Biocentrism involves two components: first, that *the Universe is finely tuned for life*, and second, that *consciousness is the key to understanding the Universe.*

Undoubtedly the concepts of consciousness and reality are closely linked, but as human beings, we tend to think that there are two worlds, a world that is outside and a private world that is inside our heads; that is, seemingly, there is an external and internal reality. However, external reality only exists as a function of our consciousness; the only thing we can perceive as human beings are our perceptions, and nothing exists outside our consciousness. The "outside" is therefore located "inside" in our mind.

Another concept expressed in the works of Lanza and Berman is that space and time are not real and are only forms that, as intelligent beings, we use to understand periods and sequences of events and

---

[54] Biocentrism: How Life and Consciousness are the Keys to Understanding the True Nature of the Universe (2010) ISBN 13: 978-1-933771-69-4.
 Beyond Biocentrism. Rethinking Time, Space, Consciousness and the Illusion of Death. (2017) ISBN 13: 978-1942952213.

places and moments where they occur. These are only conceptual and subjective elements created in our minds.

Similarly, what we see, touch, or smell, all these sensations are experienced within the mind, and none of these sensory elements are out there. All we perceive is the direct interaction of matter and energy with our minds. Therefore, anything we do not observe directly only exists in the same way as it happens in quantum physics.

Furthermore, Robert Lanza delves into this concept and tells us that the physical world around us is just an illusion created by our minds, similar to the plot of the movie Matrix. The entire Universe is located here, in my mind and consciousness, not outside; there is no such thing as "outside." Therefore, reality is not only a subjective experience of consciousness; it is created by consciousness. This concept is the essence, the core point of Biocentrism, from which reality begins and ends with the observer, as indicated by John Wheeler's Participatory Anthropic Principle. So, the cause is consciousness, and the effect is the reality, not the other way around.

My true self is the consciousness that possesses a mortal vehicle called a body; I am not a body; I have a body. Nevertheless, "the self" does not exist as an isolated entity but as part of the whole. I am not delimited outward by the epidermis; I am not the content inside a skin bag. In reality, I am, and we all are, immortal consciousnesses creators of the illusion of time and space.

The past and the future do not exist, only the eternal present. In 1955 Albert Einstein wrote shortly before his death: *"For me, the distinction between past, present, and future is only a stubborn and persistent illusion."* In the philosophy of biocentrism, there is no "beyond" either; it cannot exist if there is nothing outside of consciousness. This concept is crucial for Judeo-Christian beliefs because it dissolves the visions of heaven and hell. Even for the creed of Eastern religions, the weight of the reasoning of this ideology, to some extent, prevents reincarnation from existing since there is no one death from which to be reborn.

Biocentrism, in essence, transcends the idea of duality with a God and goes beyond determinism or concepts of randomness to finally reach the goal of locating ourselves "permanently" as an element of the whole, forming part of "Consciousness" with a capital "C."

It is having reached the perfect level of Unity that in different religions and philosophies has been named realization, enlightenment, union with God or Nirvana, and to which few humans have momentary access and very rarely permanently as perhaps Jesus Christ and Siddhartha Gautama did. The basis of the philosophy of Unity is the Hindu doctrine of Advaita Vedanta 1,300 years ago, where space, time, and even the individual are illusions. Consequently, the here and now are considered timeless and non-spatial, so the individual only exists as part of the Unity of Consciousness.

Biocentrism breaks cultural and religious paradigms and rescues the essence of being as a synonym for consciousness. It shares some concepts with Baruch Spinoza, George Berkeley, and some aspects of quantum physics, especially the mysticism that arises from it. The essential points, the icing on the cake of this theory of Biocentrism, would be: *"The mind transcends time and space as its tools and not the other way around. This concept dissolves human individuality. We should not speak of conscience but consciences in the union. Together we are all One, fused in Unity. In the consciousness of unity, the walls of time and space are imaginary, which is the prologue of immortality."*

Reality and Consciousness

After reading these surprising concepts of Biocentrism, we will briefly review other ways of observing the association between reality and consciousness using more conventional ideas. However, this does not mean they are necessarily more truthful or exclusive. Therefore, unlike the concepts expressed by Lanza and Berman, Dr. Anil K. Seth, a renowned professor of neuroscience at the University of Sussex in the United Kingdom and one of the world leaders in the investigation of consciousness, defines it as: *"the experience of the world and ourselves that we have through the brain; and our brain can only give us the best guess it can through the information sent by the sense organs, so the translation it does is a fantasy that generally but does not always coincide with reality."* In this definition, as we see, the concept of consciousness-reality is opposite to that of Biocentrism, the cause is reality, and the effect is consciousness.

This thesis of Dr. Seth is based on the ideas of the academic Alfred Korzybski (1879-1950), who, in 1931, created the famous phrase *"the map is not the territory, and it does not represent the entire territory,"* which is one of the bases of General Semantics and Neurolinguistic Programming. Korzybski claimed that human beings are limited in their knowledge by the configuration of their nervous systems and the structure of their languages.

Indeed, the only access we have to reality is through conscious experience, personal and subjective. If there is a reality beyond human consciousness, it will remain unknown since we only have access to our perceptions. The discipline in charge of studying this abstract concept is Metaphysics, which, as part of philosophy in its most general sense, examines reality's nature, constitution, and structure *(Audi, 2004)*.

Based on quantum physics and classical physics observations, a relevant question would be whether consciousness creates matter and energy or whether these are the creators of mind and consciousness *(Chopra, 2014)*.

Evidence from the two physics clearly shows that it occurs in both directions. In this way and as a consequence: "matter, energy, and consciousness are interchangeable and therefore are the same entity." Reality is perceived through consciousness, and, in turn, consciousness is the creator of reality. Therefore, consciousness is not something abstract or ethereal, and it is made up of the same elements that make up the Universe, only that we still do not know how our brain generates it from physical, chemical, and biological components. Thus, the essence of consciousness and reality is a mystery that Nature jealously guards.

The German word *Gestalt*, translated into English, means "form" and represents how we construct frames of perception of reality. All people interpret reality and make decisions based on these forms or mental and unconscious figures we create over time with life experiences. So, Gestalt theory explains how we perceive things and make decisions based on the forms we create.

Since its appearance at the beginning of the 20th century, various representatives of this psychological current created specific laws, such as proximity, similarity, closure, continuation, unity, etc. Together, they express that, in human consciousness, we perceive the elements of an image as a whole and relate them to forms known to us.

According to this theory, *figure 3.15* is an excellent example of human perception *(Jakel, 2016)*. Indeed, perception of reality as a primary element of consciousness can be falsified or distorted in many ways, for example, from ambiguous or partial sensations, as happens with illusions (visual, auditory, tactile, olfactory, and gustatory) and even more, with hallucinations derived from the effects of certain psychotropic substances or due to severe mental illnesses such as schizophrenia.

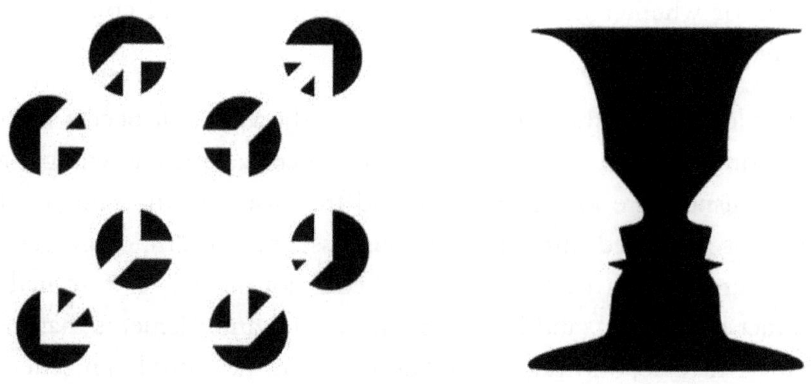

Figure. 3.15 Gestalt theory. On the left: *the intermediate edges of the cube are imaginary and generated by our brain according to the "law of closure."* ***Right:*** *Rubin's vase, an example of "multistable perception."*

Credit: Bernard Ladenthin and Alan De Smet. Public domain.

Thus, under certain conditions, reality can be detected in a misleading or erroneous way. For example, in *Figure 3.16,* we can see a drawing by the Dutch artist Maurits Escher (1898-1972), who specialized in creating images that are impossible or incompatible with reality.

The first historical record describing the misleading perception of reality is found in a passage from Rigveda in India more than 3,000 years ago. This passage is the famous parable of *"The blind men and the elephant,"* It narrates how a group of blind men wants to know through touch what an elephant is like? The conclusion reached by each one depends on the area of the animal to which they had access with their hands; thus, they describe the elephant as a log, a snake, a rope, a spear, a wall, or a fan.

At the end of the 4th century BC, Plato describes how humans can obtain false knowledge of reality from misleading visual perceptions through the *«Allegory of the Cave »* in his work "The Republic." Still, the philosopher goes further and demonstrates that we can even create paradigms from these false perceptions.

Reality is veiled or hidden from our perception if it goes against our wishes, ideas, and concepts. Prejudices and beliefs are potent mechanisms against a correct perception of reality. Thereby, magicians and conjurers know our sensory limitations as human beings and deceive our brains through well-studied and applied tricks so that we see reality the way they want us to perceive it.

Figure. 3.16 Relativity *-modified lower section of the original image-*
(M. Cornelis Escher).

Credit: Collection Gemeentemuseum - Den Haag, The Hague.
The Netherlands.
Partial and modified low-resolution images for a
legitimate use for educational purposes.

A strange phenomenon that has been observed in some people is that they can capture sensory stimuli with the organ that is not the specific one for that particular stimulus, causing perceptual fusions. This event describes, for example, the ability to see and smell sounds or perceive tastes through touch or hearing, and this quality is called *"synesthesia."* The general concept of synesthesia indicates that it is a phenomenon in which the presence of a stimulus produces additional or concomitant experiences for which there are no corresponding physical or sensory inputs *(van Leeuwen, 2015)*. However, this occurs not only with perceptual stimulation but also with emotional stimuli or those generated internally to evoke representative sensory, cognitive or affective *"synaesthetic"* experiences.

Synesthetic associations are complementary, random, and personal and are not evoked at will or chosen. However, they cannot be suppressed, are often described as automatic or unavoidable, and are stable throughout life. The current consensus is that synesthesias are not the product of imagination, nor are they metaphorical thoughts, but rather have a neurological basis.

They can occur even when one of the senses is damaged, for example, a person who can see colors when he hears words or music can continue perceiving these colors, even if he loses his sight, and this will persist for the rest of his life *(Hupé, 2015)*.

This phenomenon has been known for more than 200 years. Still, it is only since the last decade of the previous century that substantial progress has been made in its research and understanding of the mechanisms that originate it. It is known to have a worldwide prevalence of approximately 1: 2,000, much more frequently in the female sex (3-5: 1 compared to the male sex). There is a type of synesthesia called spatial localization, which is much more common in the general population ($\pm$ 10%), and this phenomenon occurs when mentioning dates or names. Many people say that when hearing or reading, for example, "year 1,200 BC". or "fourteenth century" and even the words "history or prehistory" place them spatially behind or to the left of them. The same happens when mentioning the future; when hearing something about the twenty-fifth century or a million years from now, they place it to the right or forward and further as the number is greater. Some other people even spatially locate the days of the week or the months of the year.

Others also perceive them with a specific color, in the same way that others perceive a particular color for each grapheme[55] or musical sound. Richard Feynman, whom we spoke about previously, said he saw the equations in color, and some musicians can see the sounds of musical notes with different chromatic tones, such as Olivier Messiaen *(Ward, 2013)*. Also, most likely, Mozart.

Other less common types of synesthesia include the sensation of touch with sight, described as "mirror-touch." Certain people report that when they observe someone touching another person's face, they perceive them as if somebody touched their face. Still, others include the perception of flavors when hearing or reading a word or even touching a texture.

When we talk about the Frey effect in electrophonic hearing, we have already mentioned that certain people perceive ultra-high range frequencies as odors located between 300 to 3,000 MHz. The origin of this phenomenon has genetic bases not yet fully defined, but two facts have been observed: first, there is a polygenic family inheritance pattern *(Asher, 2009; Tomson, 2011)*; that is, synesthesia is connected with various genes located on multiple chromosomes (2q24, 5q33, 6p12, 12p12 and 16q12.2—23.1) and second, that synesthesia is the consequence of alterations in the brain connectome as a result of the permanence of inter-sensory synapses present in the fetal stage and that persist during the first months of life; in fact, this phenomenon is present in almost all babies under 18 months. Neural pruning in the second year of life is likely incompletely carried out in people with

---

[55] A grapheme is a minimum and indivisible writing unit representing a phoneme (letters, numbers) or diacritical marks -commas, points, accents, etc.

synesthesia. We must remember that, at birth, the human being has 150% of the synapses it will have in the adult stage *(Huttenlocher, 1997)*.

A more recent hypothesis places synesthesia development with learning. Synaesthetic inducers, the stimuli that provoke these unusual experiences, often involve perceiving complex aspects learned in childhood, such as letters, musical notes, numbers, and months of the year. New research has shown that the associations that synaesthetic establish with these learned inducers are not random but are strongly influenced by the structure of the learned area.

Synesthesia can affect or be useful for learning. Synaesthetic children use their unusual experiences as mnemonic devices and can even use them while learning new abstract concepts *(Watson, 2014)*. In the same way, it has been identified that there is an increase in the incidence of synesthesia in those children forced to learn one or more languages in addition to their mother tongue at an early age *(Watson, 2017)*.

In the images obtained employing diffusion tensor MRI of the brains of synaesthetes, subtle differences have been found in the density of gray and white matter in certain parieto-occipital and frontal regions compared to non-synaesthetic people. Synesthesia also occurs in post-childhood stages, and these originate in two ways: first, when a sense organ is lost, especially sight, visual-tactile, and visual-auditory synesthesias are acquired. Second, when there is a hearing loss, auditory-tactile synesthesia frequently appears. The second form also occurs through a hallucinogen like LSD (lysergic acid diethylamide), which triggers this phenomenon in addition to hallucinations.

Those with synesthesia perceive reality in a different way from the rest of the population. When the brain is still immature, it requires redundant elements to learn and understand its environment and includes all the physical elements available in time and space that can stimulate each sense organ. Although it is not synesthetic in the adult age, something of this structure remains to help identify reality.

Wolfgang Köhler, a psychologist at the University of Berlin and one of the leading theorists of the Gestalt School, developed a test using a pair of images *(Fig. 3.17)*. First, he assigned a name to each one *(Bouba and Kiki)*. Then, the figures are shown to anyone and asked to guess which is which. As a result, 97% of people assign *Bouba* to the purple rounded figure and *Kiki* to the orange angular figure. This test shows that humans do not name objects arbitrarily.

Figure. 3.17   Bouba and Kiki.
Wolfgang Köhler and Vilayanur S. Ramachandran.
Credit: Andrew Dunn. GNU Free Documentation License.

Closely related to reality and consciousness, a controversial new theory has been developed indicating that our brain is genetically endowed with an innate virtual reality generator that, through experience, becomes a creative or predictive model[56].

---

[56] This generator comprises qualions, which are virtual composites of bosons and endogenous electromagnetic fields. They are the theoretical constituents of consciousness or qualia.

This model works most intensely during rapid eye movement (REM or REM) sleep and provides optimal conscious waking experiences. This statement is based on several observations in functional magnetic resonance imaging studies that support the hypothesis that "awake" consciousness emerges from REM sleep ("asleep" consciousness), and dreaming lays the foundation for perception during wakefulness. Thus, sleep plays an essential role[57] in maintaining and enhancing the ability to shape the world.

So, the basis of consciousness relies on the inference process from the generation of virtual realities (both during sleep and wakefulness). The brain chooses the one or ones that best apply to the sensation received according to previous experiences. Thus, the awake brain relates to the world by predicting the causes of sensations. The brain is actively purified during sleep to generate more efficient predictions during wakefulness.

Our perception of reality in the wakeful state is then based on reconstructions of the world by the brain, which are fantasies or hypotheses that explain the sensations we receive from the environment. This ability to shape the world underlies the perception level not only during wakefulness but also in sleep. Thus, dreaming is an essential prelude to wakefulness reality perception *(Hobson, 2014)*.

In another context, there are certain circumstances where the perception of reality is altered, such as in the case of events where there is a dissociation between consciousness and attention. Although consciousness and attention seem to be closely connected (mindful attention), there are processes in which one component may be present in the absence of the other.

The dissociation between attention and consciousness occurs because overlapping neural networks in the brain carry out different brain processes. The brain allows us to pay attention to something

---

[57] Sleep deprivation is associated with severe perceptual disturbances and even leads to the development of psychotic behaviors.

without being fully aware of it; this is the basis of *subliminal perception*. This perception occurs through imperceptible stimuli or messages due to their brevity and low intensity. However, since the beginning of the 20th century, it was already affirmed that it had a certain relationship with behavior and decision-making.

Regarding this point, its fundamental importance has been much debated, among other things, for having tried to use it since the fifties for commercial and political purposes, specifically to influence the preferences of consumers and voters. However, a systematic review has recently been conducted that confirms that unconscious attention influences choice judgment, but in general, it is temporary and of low intensity *(Newell, 2014; Arese, 2019)*.

The problem of precisely defining the correlation between attention and the conscious or unconscious occurs because, during experimentation, it is practically impossible to separate the threshold of subliminal perception from the supraliminal. Therefore, during the research process, answers are being analyzed on the border between the conscious and the unconscious. A typical example of this disaggregation without the need for subliminal messages would be the decision when making an impulse purchase; this is done at a level where attention predominates over consciousness, or the process of hypnosis in which attention is focused on the voice of the hypnotist without awareness of the implicit suggestion.

There is also the inverse of the dissociation between consciousness and attention, which would be conscious-inattention. A typical process in everyday life includes automatic behavior and habits; for example, I am sitting down having breakfast and reading some exciting news on my tablet. I am aware that I am eating, but I have little or no idea what I am eating or how much I have eaten because I am focusing on reading the news. Another, more frequent example is driving a car while simultaneously planning my activities for the day or answering a call on my cell phone.

All these behaviors or automatisms depend on both implicit and explicit memory that work interactively. Indeed, it is well known that consciousness is closely related to memory. Memory, as a monitor of learning, presents past experiences to consciousness so that it can compare them with current reality and thus decide if they have correlation and significance or not. If it does, then memory will show them to the intelligence that will be able to choose the most appropriate response *(Fig. 3.19)*. However, memory is generally quite fallible in the short and long term, except for intense experiences or those involved with survival. We are, for example, unable to faithfully narrate our past and can only relate some significant memories and events with relative precision in time, nuancing them with emotions and frequently with errors or additions to which we resort to make sense of our evocation and, thus, fill in the gaps of memory.

Our mind is not a video camera; it does not work like that. The autobiographical process, for example, is a task that requires effort and the support of people with whom we shared the past, and in that way, we improve our memory. Our history, at last, will be not how it happened but how we remember it, and frequently it becomes the only reality that existed. However, as we will see later, there is a much broader unconscious memory where many memories are registered, which we cannot consciously evoke. Still, we have access under certain circumstances or by specific psychoanalytic techniques.

Cybernetic Consciousness

Some current knowledge regarding cognition and consciousness is based on artificial intelligence concepts. They explain its origins and evolution concretely and correctly, with elements such as the Markov blanket[58] or Bayesian networks that will not be included in this chapter because it would take us away from its purpose.

The first attempts to create computational models of artificial consciousness began in the late 1990s. Two decades later, some researchers argue it will be achievable once short-term engineering challenges are overcome *(Reggia, 2013)*.

Others, however, think that consciousness cannot be implemented in machines *(Haladjian, 2016)*. Although there is the ability to program behavior based on rules and machine learning, it will never be possible to reproduce emotions or empathy through the programming of control systems so that they will be only mere simulations. The arguments that suggest the impossibility of achieving these include considerations about evolution, the neuropsychological aspects of emotions, and the dissociation between consciousness and attention.

---

[58] The Markov blanket comprises bounded subsets that maintain statistical independence from the rest of the network. Translating this mathematical concept to the nervous system, for example, cognitive capacity is based on groups of neurons that have a more significant association among themselves than with other neuronal groups; these packets together form models that allow the perception of objects and sensations their comprehension. This concept, therefore, also has application to explain intelligence and partially consciousness *(Yufik, 2016)*.

Typical of human beings, which we have already explained, but it simply means that: *"not everything that seems is, and not everything is what it seems."*

Another theory that explains why it is unlikely to create artificial consciousness is the "neuronal group selection theory," which indicates that the brain is a selection system, not an instructional system like a computer. During the development and behavior of an organism, large numbers of variant neural networks are generated. These constitute complex repertoires from which the circuits formed based on value systems are selected to ensure the adaptive behavior of the organism *(Seth, 2006).*

Biologically, consciousness arises from a vast, highly complex network called the interactome, which is the total of molecular and genetic interactions in a given cell. If this interactome manifests itself in billions of neurons through their intercellular links, they are the origin of all brain functions. This system of connections in the nervous system is known as the *connectome*, which is a tangled communication map in the brain. If we ever build an artificial network with this type of structure, the resulting cybernetic entity is likely to be conscious as well.

In a recent publication *(Hildt, 2019),* the capacity that we humans have to realize if a computer has consciousness is questioned. An adequately programmed computer is a mind; it has programs that understand and therefore has primary cognitive states. While we know about human consciousness from the first-person perspective, artificial consciousness will only be accessible to us from the third-person perspective. The consciousness of machines with increasing artificial intelligence is or will be precisely that, a machine consciousness, different from ours or any other living being.

Conscious and Unconscious Intelligence

With this topic, we have finally reached the peak of the evolutionary processes that began billions of years ago with the birth of our Universe and later gave rise to the first living microscopic entity, "our first ancestor." From the moment life appeared, it has required to persist various strategies based on variability. As the Ukrainian geneticist Theodosius Dobzhansky wrote, *"Nothing in biology makes sense except in the light of evolution."* Based on this statement, two new concepts have recently been coined around the origin of life: the first says that the purpose of life itself is evolution, and a derivative of this, the second indicates that the evolution of the inorganic world was the one that gave rise to life. Thus, the order of the traditional paradigm is the other way around; first, it was evolution and then life. Therefore, life is a consequence of evolution *(Takeuchi, 2017).*

Higher consciousness and intelligence result from many evolutionary forms of natural experimentation used to ensure survival. This experiment has led us not only to exist and be the dominant animal species, capable of carrying out the most beautiful or heinous acts as a result of a complicated mind that phylogenetically has had to evolve according to the challenges that the environment has been presenting over millions of years. In this way, the brain has grown not only in size but mainly in complexity.

From this observation, the *"Trinitarian or Triune Brain Theory"* emerged about 60 years ago, conceived by Dr. Paul MacLean (1913-2007), a researcher at Yale University.

This theory is based on observations made in comparative anatomy between the human brain and that of other animal species *(Fig. 3.18)*. Although currently, both the theory and its conclusions have been proven that they are not valid in their entirety and therefore are no longer accepted in the scientific community, instead, scientists agree that it is a good analogy and a way to explain why we have specific capacities and behaviors[59] from the evolutional point of view.

The theory says that the human brain is a superposition of archaic brains having in the center the *Reptilian Complex* or reptilian brain that includes: the brainstem, the cerebellum, and the basal ganglia. MacLean proposed that the reptilian complex is responsible for instinctual behaviors typical of species in which aggression, dominance, and territoriality prevail. Next is the *Paleo-mammalian Complex,* which includes the limbic system from which the motivations and emotions arise, as well as mating and protective behaviors of parents towards their young; however, it is not exclusive to mammals.

Finally, the *Neo-mammalian Complex* is made up of the neocortex, a structure from which MacLean considered to be the most recent step in the evolution of the mammalian brain, mainly due to the development of the prefrontal cortex, which gives it the possibility of possessing superior brain functions such as the capacity for abstraction and creativity, the potential to develop a sophisticated language, to plan and systematically solve problems, and to have complete self-awareness, therefore exclusive of the human being *(MacLean, 1990)*.

Now, in the evolved and complex human brain, it is evident that intelligence and consciousness are closely linked to such a degree that, in general, there is a proportionality between the two. Nevertheless, intelligence can exercise its function in well-defined situations on unconscious grounds.

---

[59]   It is currently recognized that the MacLean concentric evolutionary system is imprecise since it is based on a linear evolution. It is much more complex and adapts much better to the branched evolution model. In his book *"The Dragons of Eden,"* written in 1977, Carl Sagan extensively references the theory.

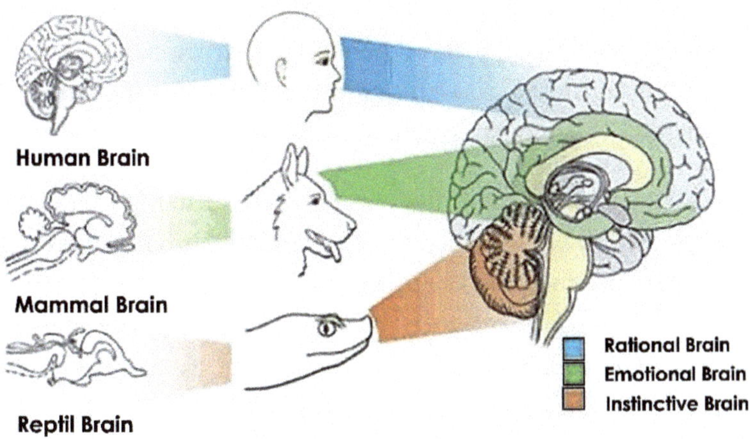

Figure 3.18   The Triune Brain. Paul McLean.

Credit: Public Domain.

This situation starts from consciousness being only a tiny portion of the psychic system. This evidence was the main foundation that Sigmund Freud (1856-1939) used to develop his Theory of Psychoanalysis, which describes that an enormous amount of information is registered in the unconscious, the entry of which has occurred through sensations and perception. Since the unconscious territory is much broader than the conscious, it also houses a much larger unconscious memory that has no place in conscious memory or is repressed due to conflictive situations.

It is where a large part of our experiences is located. To access it methodically, Freud created a series of techniques that include —free association, transference, and interpretation of speech, especially in relation to dreams and failed acts *(Restrepo, 2015)*.

Freud described three levels of consciousness: the first level is occupied by *«the conscious, »* which refers to perception through the sense organs closely related to reality, here and now. The second level, *«the preconscious, »* consists of feelings, thoughts, and fantasies that are not permanently present in consciousness but can appear at any time through evocation. Finally, the third level, *«the unconscious, »* is where all the contents, emotions, ideas, and conflicts that we have repressed and forgotten are registered due, among other things, to the suffering that having them in consciousness would cause us.

We can eventually remember most of these items stored there in three ways: first, by re-experiencing the same disturbing events; second, through certain situations or circumstances that may or may not be related, but that somehow give us access to that past reality that we have forgotten. Finally, in third place, it is precisely through psychoanalysis. The latter has the advantage of having access to the unconscious in a relatively complete and structured way, allowing the necessary psychotherapy simultaneously *(Freud, 1940).*

Like memories, intelligence can manifest itself both consciously and unconsciously. How it displays consciously is the most important as intelligence and consciousness are fed back continuously with information. Consciousness shows the context of the perception of reality so that intelligence can choose and decide responses or actions.

In *Fig. 3.19,* we can graphically observe the intimate relationship between —*Consciousness and Intelligence*— Consciousness presents scenarios of the present and the immediate past through short-term memory. It also does so with the experiences of ancient history registered in long-term memory, specifically by the so-called «Explicit Memory, » also called *declarative memory.*

This memory is clearly associated with consciousness or at least with conscious perception *(Coge, 2013).* Retaining its contents depends on the hippocampus as well as the central areas of the brain's temporal lobes. For its part, «Implicit Memory, » which is also part of long-term memory, is independent of consciousness and is called *non-declarative* or *procedural* memory. This memory includes abilities, habits, and conditioned reflexes. The retention of information does not involve the hippocampus, but it does the caudate nucleus and some regions of the occipital lobes and the cerebellum *(Carrillo-Mora, 2010).*

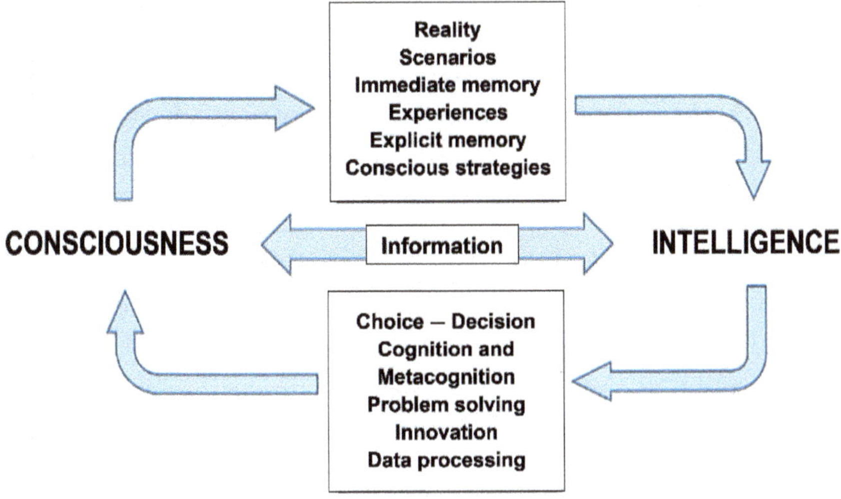

Figure. 3.19   Components of the interrelation between
Consciousness and Intelligence.

Credit: Own Original Work.

This memory is associated with consciousness or at least with conscious perception *(Coge, 2013)*. Retaining its contents depends on the hippocampus and the central areas of the brain's temporal lobes. For its part, «Implicit Memory, » which is also part of long-term memory, is independent of consciousness and is called *non-declarative* or *procedural* memory. This memory includes abilities, habits, and conditioned reflexes. The retention of information does not involve the hippocampus but the caudate nucleus and some regions of the occipital lobes and the cerebellum *(Carrillo-Mora, 2010)*.

Some elements contained in explicit memory can become components of implicit memory with learning, primarily through repetition of many trials, for example, "riding a bicycle." Recent findings through functional magnetic resonance imaging conclude that implicit and explicit memory systems collaborate dynamically to perform specific tasks, a situation that disappears in people with dementia, such as —Alzheimer's disease—, who lose a large part of their explicit memory, leaving the implicit one relatively well preserved *(Koenig, 2008)*. Implicit memory facilitates behavior without being aware of prior learning processes or resulting knowledge. That is why most people who have dementia engage in inappropriate behaviors. On the other hand, the explicit memory dependent on consciousness is essential for conserving learned behaviors.

Now, regarding *Short-term Memory*, it can keep experiences in mind for only a short time, and there are two types: «*Immediate memory*» that covers only a few seconds; however, it involves all modalities (visual, auditory, and tactile among others), the most important thing about this memory is that it gives us *"the continuous sense of the present."*

The second type of short-term memory is «*Working Memory,* » which can hold experiences in mind long enough to carry out sequential actions and involves sustained attention. The fundamental difference with immediate memory is that while in this one, we use a relatively simple neural system, working memory activates multiple sites in the brain where information is temporarily stored; that is why it is also called —*active memory*. With this memory, we are unaware of all the data used simultaneously, as happens when driving a car. In this activity, we launch different sensory and motor information that is processed simultaneously. All this information is located in different brain regions (primary sensory areas of the prefrontal lobe and the dorsal-median nucleus of the thalamus, among others). We pay attention to them according to our needs at the time of their application.

Working memory is, composed of two subsystems: one for the control of attention called the *central administrator,* which is the one that coordinates the systems in charge of information from *"language;"* and another called the *visuospatial agenda,* which controls how *"mental images"* are processed. In addition, some contents of working memory have the possibility of accessing long-term memory, which is why you never forget how to ride a bicycle. This long-term memory, as we have already said, is divided into explicit (conscious) and implicit (unconscious) *(Solís, 2009)*.

"Intelligence" responds to the information provided by consciousness through cognition and metacognition, which carry out data processing, choosing the best alternative in solving problems, either by innovating or using resources previously recorded in long-term memory, especially explicit memory. As we can see, the two functions are closely linked, and many of the components correspond to both faculties, such as conscious strategies and cognition. With the information received, smarter people can produce more meaningful connections or elaborations during learning and data coding processes than people with lower intellectual abilities. So, brilliant people differ from those who only have much knowledge; they can both have the same amount of information, but the difference lies in what they do with it.

Now, intelligence on a conscious basis is different from unconscious intelligence. Some authors hardly relate intelligence to unconscious memory and consider that the interdependence between consciousness and intelligence is essential *(Joordens, 2013)*. However, as we shall see, phylogenetically, the first terrain on which primitive intelligence acted was on unconscious grounds, and there is reliable evidence for this.

To understand how intelligence can act through the information provided by *the unconscious,* we must know that many instantaneous decisions we make are not based on conscious information. This situation depends on emotional and motivational systems that constitute cognition and the unconscious mind. Quick decisions depend on small pieces of information or experiences located in the unconscious perception and the implicit memory, through which we decide the immediate response or behavior to follow. Let us consider the example of a football player who is about to receive the ball while running; he needs to keep the angle of his eyes on the ball while deciding how he will catch it and, at the same time, how he will outwit the two opponents in front of him.

These strategies are called *«heuristics, »* through which a complicated problem is solved in a relatively simple way *(Isenman, 2013)*. In the player's case, he makes use of his heuristic vision.

Another homologous strategy is the so-called heuristic identification, which we can better explain through another example: suppose that we are walking through a forest, and suddenly out of the sideways of our eye, close to us, we observe an elongated image in the shape of an "S" on the ground. This image makes us instantly jump and escape that place as quickly as possible. We decided to move out unconsciously since our implicit and ancestral memory told us it was a snake. Later, we can consciously approach it again only to discover that it was a tree branch. Thus, conscious actions involve more widespread activations in the brain than similar but unconscious ones. The time required for a stimulus to reach consciousness is between 300 and 500 milliseconds *(Ortinski, 2004)*.

These heuristics depend on our species' evolutionary heritage inherited over hundreds of thousands of years, just as it happened when we hunted prey or fled from a predator. Intelligence on genetically transmitted unconscious bases results in *innate behavior*, which is the most important for most animals; its survival depends on it, which we call *"instinct."* Therefore, the unconscious intelligence is much older than the conscious one, and thanks to it, in large part, human beings have managed to survive.

You should note that the heuristic in the case of the soccer player requires the use of superior intelligence compared to the example of the snake, which is merely an ancestral reflection. The anatomical place for the unconscious perception in humans is registered in the cerebral tonsils, particularly the *right* one *(Mormann, 2011)*.

Although there is no specific time sensory stimulation for intelligence to act on exclusively unconscious bases, various tests have shown that it is generally within a few tenths of a second. The words "fast and scarce" go well with the information that the unconscious gives to the intelligence, and it is through actual mental shortcuts that the most critical decisions are made in the blink of an eye. In this case, the statement that "less is more" becomes true.

Human beings often act based on incomplete information, which is the foundation of intuition and our usually accurate hunches. But, as the renowned German psychologist Gerd Gigerenzer states, we own an extensive heuristic repertoire that he calls an *"adaptive toolbox."* This repertoire has millions of years of evolution, so it is wholly perfected, especially with life and death decisions. On many occasions, we have made better decisions based on hunches throughout our lives than when we stop to reflect. It has been shown, for example, that financial analysts do not predict markets better than partially informed amateurs who only use simple heuristics *(Gigerenzer, 2008)*. Intuition, insight, instinct, smell, presentiment, or hunch are many words to describe the same phenomenon, which indicates its great importance in our lives.

Since the middle of the last century, heuristics has become a technique or strategy used systematically to solve certain research problems and create algorithms. However, choices based exclusively on intuition can also lead to errors because the construction or ideation of solutions through heuristics often starts from reasoning by analogy. Therefore, based on the available empirical evidence, it is not highly recommended to use heuristic methods systematically in scientific research *(Hilbig, 2010)*. As we already mentioned, unconscious intelligence works best when information is scarce. Nevertheless, even with complete information, the human mind ignores much of it and automatically selects it in order of importance. —For example, 1n 7h3 n3x7 p4grar4ph i7 d03s n07 ma773r 7h3 0dr3r 0f 7h3 L377r3s 1n 3ach wr0d 0r 1f 7h3y ar3 NUM83R5, y0u c4n r34d 7heM W1ht0u7 4 L07 0f 3ffr07 [60].

Heuristics play an essential role in unconscious understanding as it selects the most familiar option to us and discards most of the information. Thus, the cognitive abilities of the unconscious mind are highly integrated with the conscious mind, but they are faster because they involve fewer complex connections and are closely related to

---

[60] In the next paragraph, it does not matter the Order Of the Letters in each wOrd, Or if they are NUMBERS, yOu can read theM Without a Lot of effort.

survival. In this way, the conscious and unconscious mind is a continuous complex without defined borders, understanding by mind the totality of the cognitive faculties that include thought, perception, consciousness, judgment, memory, and intuition.

Unconscious memory is so old that it has had enough time to connect different species through evolution, so vital information can be transmitted through genetic codes and create a collective unconscious memory. However, most instinctual behaviors are species-specific because they involve particular adaptive purposes.

Ethology is the branch of biology and experimental psychology that studies the *innate behavior* or "instinct" and the *learned behavior* of animals in their natural environment. However, we can locate human behavior parallels with other mammals through comparative psychology, especially with other hominids. Thus, in the example of the snake, we said that we react by putting distance instantly because our "ancestral" memory indicates the danger of being close to that mental image. That reaction happens in most mammals, but not so, for example, in predatory birds in which, on the contrary, it awakens in them the hunting instinct.

Gene Ezia Robinson, the foremost American entomologist, is a pioneer in applying genomics to studying social behavior. He points out that, to be understood, animal instincts must be seen as *Matryoshkas dolls*; instincts nest one inside the other. For example, the smallest doll corresponds to the oldest instinct from which the next one is molded depending on the new adaptation needs. This mechanism often involves increasing the complexity of gene sequences, thus modifying the neuronal genome through mutations and giving rise to new, increasingly complex brain circuits *(Kapheim, 2015)*. On other opportunities, brain plasticity without a mutation, which conditions changes through modifications in gene expression.

This plasticity is closely related to Epigenetics, which occurs through *DNA methylation, histone modification, and small non-coding RNAs*. In none of the three mechanisms, there is any modification in the sequence of genes within DNA, only their expression changes. When these modifications occur in the hippocampus, animals acquire the ability to memorize any transcendent experience that will translate over time into new responses and behaviors.

The evolutionary changes of these epigenetic mechanisms will shape the learned behavior, converting it over time into an instinct that will be stably inherited for several generations until, with time, environmental changes modify the need for this behavior in favor of another, a new that means a greater chance of survival *(Robinson, 2017)*.

For example, in the specific case of wolves, tens of thousands of years ago, only those who could produce fewer stress hormones such as cortisol and control their aggressive behavior towards humans were able to self-domesticate. Thus, in a docile and peaceful way, they established greater contact, thus obtaining food from the remains left by the men in a simple and safe way. Wolves ancestrally already possessed three instinctual abilities that facilitated their «self-domestication. » These are: cooperativism, tolerance, and attention, hence the second stage of their domestication was enabled, but now, it was the man who carried it out with the descendants of those self-domesticated wolves, which had to make new behavioral changes to accept becoming helpers for hunting, for transportation —for example, pulling sleds— and finally even for grazing, avoiding attacking animals that generations ago were their prey and now, defending them from the threat of other predators. Definitely, the ancestors of dogs had the potential to become the best pets *(Range, 2015)*.

The memory of the collective unconscious is the animal evolutionary genetic source of ancestral innate behaviors. It is undoubtedly a primordial element for humans since it allows us to understand a large part of the reason for our feelings, attitudes, and behaviors related to other animals, especially primates. However, it is the content of our *"collective human un-conscious memory"* that specifically possesses the history and nature of *Homo sapiens* that emerged 200,000 years ago. In this particular collective unconscious is inscribed, in addition to our instinctual behavior, the entire cultural, psychological, and spiritual heritage, where our beliefs, fears, achievements, creations, and desires, engendered over millennia, have been shaped. The prominent Swiss psychiatrist Carl Gustav Jung developed a whole psychological theory of this collective human unconscious which he claimed manifests itself through symbols and signs that he called archetypes.

Well, so far, we have finished analyzing the interrelation between consciousness, intelligence, and memory. We will now continue to examine consciousness from the philosophical point of view known as Universal Consciousness.

Universal Consciousness

Universal Consciousness or Panpsychism is one of the oldest thoughts, whose bases probably go back to the very origin of Homo sapiens. It arises as an extension of the social attribution of consciences typical of many animal species for millions of years (already mentioned when we reviewed animism within the evolutionary theories of consciousness and Biocentrism). Furthermore, it is related to a holistic vision that explains that the entire Universe is an organism possessing a cosmic and ubiquitous consciousness.

This ideology confers mental qualities to all elements of the physical world, from its fundamental components, the atomic particles, which are coherently combined to generate higher levels of perception and experience as they integrate into life and manifest on an ascending scale of complexity.

Beginning with viruses and continuing with the cells, later with plants, fungi, animals, and finally, human beings on Earth.

This philosophical theory was embodied in the 7th century BC. by Thales of Miletus, developed by great philosophers and thinkers throughout history such as Plato, Heraclitus, and Giordano Bruno, and in the seventeenth century by rationalist philosophers, Spinoza and Leibniz helped develop the concept of Universal Consciousness. But it was in the nineteenth century that panpsychism reached its maximum expression through philosophers Schopenhauer, Royce, William James, and the psychologists Fechner, Wundt, and Lotze.

However, in the 20th century, most intellectuals rejected this theory when the new cosmovision represented by logical positivism and empiricism[61] was created. In these theories of knowledge, only verifiable statements, whether logically or empirically, are meaningful; that is, they imply the scientific method as a starting point. Nevertheless, this situation continues to prevail in the 21st century; however, there is greater openness given that for several years it has been fully demonstrated that certain events are only observable and non-reproducible, from which disciplines such as statistics and probabilistic arise although logical positivism and empiricism are still prevalent in the 21st century, there has been an increase in openness to Universal Consciousness since it has been fully demonstrated that certain events are only observable and not reproducible.

Historically in Eastern culture, Advaita Vedanta, an 8th-century Hindu doctrine, asserts that *"Consciousness is the basis of the Universe and as such the Universe is permeated by Consciousness" (Kafatos, 2016)*. This doctrine, as well as Shaivism, are non-dualistic philosophies; they are monistic and show the perspective of the individual, the Universe, and the nature of consciousness, all as a Unit. The underlying premise is that the human being as an individual reflects the *"fundamental consciousness,"* and the entire Universe is reflected in each individual; therefore, *the Universe is consciousness.*

One of the Universal Consciousness bases is that there is no life without consciousness. Each cell part of an organism has a rudimentary consciousness, typical of its environment, which allows it to perform joint functions. Thus, there are scales of consciousness in Nature, and each living entity has its level according to its ability to perceive its environment.

Apparently, only the most evolved also have the power to perceive themselves. However, several contemporary authors who defend the philosophy of Universal Consciousness mention that if each cell is

---

[61] Empiricism is a philosophical theory stating that knowledge comes only from experience and observation of facts.

capable of self-organizing according to its individual needs, they wonder, is this done deterministically, or does it include primitive self-awareness? *(Niskama, 2015)*.

Simple deductive reasoning derived from this doctrine would be that the Universe is the creator of life, and life generates consciousness. Therefore, the whole Universe is the source of consciousness because it is inherent to nature. This simple syllogism discovers one of the most beautiful truths through which profound millenary philosophical concepts and current knowledge of quantum physics are united: *"Consciousness and the Universe are the same entity since there is a Consciousness behind the reality of the Universe."* The ancient oriental beliefs now become verifiable concepts in quantum physics experimentation laboratories, where it is confirmed that the experimenter's consciousness can create matter, energy, and reality. Modern Biocentrism, in turn, indicates that the Philosophy of Unity is a transcendent truth and affirms that there are no individual consciousnesses but that we all form a single consciousness, the Universal Consciousness.

According to this philosophy, the "Unity" between Consciousness and Life is indissoluble. Therefore, for each individual, upon death, the end of life and their consciousness does not come; it only ends the life of a body and the brain through which that consciousness manifested, which will always be part of the Consciousness of the Whole, therefore, immortality is part of our essence.

With the topic of consciousness in its broadest sense, we end this journey that began with the formation of space-time from the so-called Big Bang, the starting point of our physical Universe, probably unique within an infinite range of the Cosmos that constitute the Multiverse and that integrate the Everything with capital "E." This, our universe, had all the necessary elements for life and consciousness to appear. Thus, we are the culmination of the Great Odyssey of our Cosmos, an adventure we probably share with many other civilizations in some remote corner of our Milky Way and other galaxies.

Epilogue

With the effort of multiple generations, we have begun to unravel the three most profound enigmas: Universe, Life, and Consciousness. The road has not been easy at all, but it is impossible to resist the challenge it imposes on us. The search for answers to these and other defiance are inherent to the essence of the human being, especially those men and women who have managed to meet their basic survival requirements and have reached the highest levels in the hierarchy of needs, such as the of recognition and self-realization[62].

In this last part, we will do a reflexive recapitulation of the main points reviewed. Let us start then.

The Universe, that immense and mysterious place we live, seems to allow us to understand it and gives us its approval to know its secrets. However, today, we do not have a clear idea of how much we know and ignore; we have probably only gotten close to a small part of its essence. That is because we have so many unanswered questions, and the most disturbing thing is that most of them are the most important, for example: How did the Universe come about? We mention the Big Bang as an almost absolute truth *(Linde, 1998)*, where the infinitesimal point that it was the Cosmos in its beginning arose from the quantum foam —whatever this means, through which space-time, matter, and

---

[62] Abraham Maslow (1908-1970) was a brilliant American psychologist and co-founder of humanistic psychology. In his «Theory of Needs, » published in 1943 under the title of "A theory of human motivation," he states the basic needs of human beings. Such as physiological ones (food, sleep, etc.), must be satisfied to aspire to the next, security, and then the next one —love and belonging— to finally arrive at those of self-esteem, recognition, and self-realization. Maslow expresses his theory through a "pyramid" in his book "Motivation and Personality" published in 1954.

energy may appear literally out of nowhere. Nevertheless, if we only reflect for a moment, a deep reflection. We simply cannot conceive that "the Whole" originated 13.8, 15, or 20 billion years ago, independently of how enormous these figures are. Even a child would ask: what was there before? So, it seems that it is essential to conceive another explanation. The Multiverse theory becomes more plausible than a single Universe, with universes continually arising. Thus, our Universe is only one among an infinite number of other universes with other laws, different dimensions, and other characteristics. There has never been a beginning, and there will never be an end. The words infinite and eternity go well with the nature of the whole, the Multiverse.

Another enormous unknown is centered around how the Cosmos is made up. With some disbelief, we ask ourselves today, how is it possible that we can only observe less than 5% of all its matter and energy despite having a "high technology" in astrophysics? The called dark matter and dark energy together make up practically the entirety of our Universe. It does not seem that we will be able to discover its intimate nature in the short term, and, for the moment, we only know its effects.

On the other hand, regarding the "Microcosm," the understanding of its structure and components is in evolution, and the standard model of particles still has enormous gaps. Such is the case of the particles that carry gravitational energy called *gravitons* and the *inflatons* of dark energy; In both cases, we are not even sure that they exist *(Tanabashi, 2018)*. In the same sense, as Richard Feynman stated, the bizarre nature of quantum physics, the pillar of the organization and dynamics of all matter and energy, « *I think I can safely say that nobody understands quantum mechanics.* »

Finally, as for the destiny of our Universe, it is also within our area of ignorance. We know that the spatial geometry of the Universe is Euclidean; that is, the Cosmos is flat, which indicates that it will expand forever. Thus, matter and energy will exist for an indeterminate time that could be trillions of years or a much shorter period, approximately

22 billion years, depending on the activity exerted by dark energy in the future.

Regarding Life and its origin, our date knowledge is a little better compared to what we know about the Universe. Although no one knows precisely where and how inanimate matter came to life, we can see a lot about the multiple scenarios and indispensable mechanisms essential for its emergence. A very recent theory involves meteorites and comets as collaborators, giving them a facilitating role for the beginning of life thanks to the transport of organic molecules, but mainly due to their ability to create favorable environments as a consequence of impacts, with the subsequent formation of hydrothermal vents, cenotes, ponds and pumice pools *(Osinski, 2020)*.

We know that Biology originated from multiple natural experiments with lipoprotein polymers and nucleic acids in different environments; some of these tests worked and were able to progress and self-replicate, thus, protocells evolved to become primitive cells. Finally, one of them persisted due to various random factors, becoming the first ancestor of all life on our planet. Currently, a very popular question is whether viruses were actually the first forms of life.

Our world is the only place where we can be sure that life exists. However, the conditions that allowed it to develop may have been repeated or will be repeated on some of the trillions of planets that exist in the Universe. Finding life in other star systems in our galaxy is a daunting task that will probably require many years in the future, especially if we want to find intelligent life. The conditions for developing unicellular life are much simpler than for complex life and even more so with the capacity to create science and technology.

There is a philosophical hypothesis that describes that life is universal, and it is called *"Hylozoism"* (from the Greek ὕλη ~hyle, matter, and -zwo- zoo, animal), which asserts that all matter is alive *(Rucker, 2008)*. The concept dates back to the school of pre-Socratic philosophers. This line of thought is similar, although with specific differences with the philosophy of universal consciousness, and in both cases, no one can verify them.

The Gaia hypothesis derived from hylozoism was coined by the chemist James Lovelock in 1969, who asserted that the atmosphere and the superficial part of the continental and oceanic crust behave as a self-regulating system that favors adequate conditions for the maintenance of the Biosphere. This philosophy is based on the fact that the environment and life evolved in a linked way and mutually supported each other. This hypothesis, in general terms, is quite correct, as long as it does not include supernatural components in theory.

Finally, we come to the ability of the living beings to perceive their environment and self-perceived what we call Consciousness. Anatomically, in humans, it is located in the brain in a wide area called the Neural Correlation of Consciousness, constituted by the thalamus, the gateway to all sensory input. Then, through complex pathways, the information is carried to the entire cerebral cortex, where the phenomena of cognition and consciousness are created. Although we know that the transmission of information between neurons is through electrical and chemical synapses, we essentially ignore how the biological and quantum phenomena give origin to perception and the phenomenon of consciousness; this is an essential point whose nature is still an enigma. However, we understand enough of this phenomenon so that one of the goals of cybernetics in the next 20 to 30 years would be to create a machine with artificial consciousness based on the *connectome,* the model of human neural consciousness. This goal is planned to be achieved using both digital and quantum computing.

A concurrent and generally proportional element to Consciousness is Intelligence. These two components of the mind interact through the information mutually facilitated, allowing the entity that possesses them to make appropriate decisions according to the specific stimuli it perceives. Nevertheless, even intelligence on unconscious bases is able to carry out immediate responses based on heuristics, which initially appeared as a means of survival. Now we know it is one of the main foundations of innate behavior. We, humans, use heuristics not only in a simple, instinctive way but also in a complex way due to our degree of intelligence, and we identify much of it as intuition, sagacity, or insight. It is even used to carry out scientific research.

Throughout these pages, we have been aware that currently, our knowledge on basic essential issues but of extreme importance is very limited. However, we have a considerable excuse: humanity made its debut in this world just 200,000 years ago, corresponding to only 0.005% of the time since life appeared. There are so many unsolved questions and so many areas for research that have barely been touched that humanity in the future will be very busy with the task of solving them. We will have the opportunity to expand human knowledge in a way never seen before and at unimaginable levels thanks to a new tool that is now in its infancy but which will be decisive and permanent in a few years, I mean Artificial Intelligence (AI). Because it is not biological, this intelligence will have learning algorithms that will be very different from humans, with capacities for induction, deduction, and application of knowledge superior to the brightest human minds. Hence the applications of their knowledge will allow the creation of technology and the development of science at breakneck speeds.

These events that will happen in a few decades have been called *"technological singularity."* This event of permanent changes throughout the world will indeed happen before the year 2050, and jointly transhumanism will be developed as a philosophy that indicates that the human being will be able to improve himself through science and technology in terms of genetic, structural, and functional aspects. These changes will most likely be based on genetic engineering and the fusion of biology with cybernetics, having nanotechnology as an interface. If this perspective is consummated, transhumanist intellectuals mention that the future society will be made up of —posthuman beings.

To conclude, it will be necessary to mention that accepting these changes in the immediate future should be carried out cautiously due to the potential dangers they represent.

In 2014 the great physicist Stephen Hawking stated: *"The success in the creation of artificial intelligence could be the greatest event in the history of humankind. Unfortunately, it would also be the last unless we learn how to avoid risks."* If we can maintain control over this new intelligence, we must be optimistic as it will mean significant benefits and unprecedented scientific and technological progress for humanity. There will also be the possibility of unraveling great mysteries, including the essence of the Universe, Life, and Consciousness.

"The future has many names. For the weak is the unreachable.
For the fearful, the unknown. But, for the brave, it is the opportunity".

Victor Hugo.

Glossary

Abhidharma
They are ancient Buddhist texts (3rd century BC and later) containing detailed scholastic presentations of the doctrinal material that appears in the Buddhist sutras.

Acytota
Set of all living beings that are cellular.,

DNA
Deoxyribonucleic acid.

Advaita Vedanta
A non-dualistic branch of Hinduism affirms the unity between souls (atman) and divinity (Brahma).

Anthropic
Relative to human beings. Produced or modified by human activity. In Cosmology, the anthropic principle refers to the theory of the universe and the reality of human existence.

Biology
Natural science studies life and living organisms, including their physical structure, chemical processes, molecular interactions, physiological mechanisms, development and evolution.

Cancer
The term cancer encompasses a large group of diseases characterized by the development of abnormal cells, which divide, grow, invade and spread uncontrollably in any part of the body.

Chaos
Seemingly erratic and unpredictable behavior of some deterministic dynamical systems with great sensitivity to initial conditions.

CERN
Conseil Européen pour la Recherche Nucléaire.,

Cytota
Set of all living beings that are cellular,

Sewer
It is the posterior orifice that serves as the sole opening for the digestive, reproductive, and urinary tracts of many vertebrate animals.

Cognitive
Belonging to or relating to knowledge.,

Constant
It is that magnitude whose value does not vary over time.

Constellation
It is an area on the celestial sphere in which a group of visible stars forms a perceived outline or pattern, typically representing an animal, a mythological person or creature, or an inanimate object.,

Coursera
An American company that specializes in delivering massive open online courses.,

Chronology
A science that aims to determine the order and dates of historical events.,

Quasar
Quasi-stellar radio source acronym.,

Abhidharma
They are ancient Buddhist texts (3rd century BC and later) containing detailed scholastic presentations of the doctrinal material that appears in the Buddhist sutras.

Acytota
Set of all living beings that are cellular.,

DNA
Deoxyribonucleic acid.

Advaita Vedanta
A non-dualistic branch of Hinduism affirms the unity between souls (atman) and divinity (Brahma).

Anthropic
Relative to human beings. Produced or modified by human activity. In Cosmology, the anthropic principle refers to the theory of the universe and the reality of human existence.

Biology
Natural science studies life and living organisms, including their physical structure, chemical processes, molecular interactions, physiological mechanisms, development and evolution.

Cancer
The term cancer encompasses a large group of diseases characterized by the development of abnormal cells, which divide, grow, invade and spread uncontrollably in any part of the body.

Chaos
Seemingly erratic and unpredictable behavior of some deterministic dynamical systems with great sensitivity to initial conditions.

CERN
Conseil Européen pour la Recherche Nucléaire.,

Cytota
Set of all living beings that are cellular,

Sewer
It is the posterior orifice that serves as the sole opening for the digestive, reproductive, and urinary tracts of many vertebrate animals.

Cognitive
Belonging to or relating to knowledge.,

Constant
It is that magnitude whose value does not vary over time.

Constellation
It is an area on the celestial sphere in which a group of visible stars forms a perceived outline or pattern, typically representing an animal, a mythological person or creature, or an inanimate object.,

Coursera
An American company that specializes in delivering massive open online courses.,

Chronology
A science that aims to determine the order and dates of historical events.,

Quasar
Quasi-stellar radio source acronym.,

ISO
International Organization for Standardization. ISO is a setting that allows you to control how bright or dark a photo is and is a standard that tells you how sensitive the film / digital sensor is to light. The higher the ISO value, the brighter the images will look.

ΛCDM
Cosmological Constant (Lambda) - Cold Dark Matter.

LASER
From the English acronym LASER, light amplification by stimulated emission of radiation.

LUCA
Last Universal Common Ancestor. It is the most recent common ancestor of all life on Earth today.

Memory
Brain function allows the body to encode, store and retrieve information from the past.

Methylation
It is the addition of a methyl group (-CH3) to a molecule.

Microbiota
The normal microbiota or microbiome is the set of microorganisms located in a usual way in different places in the bodies of multicellular living beings., 98, 99, 108

Microwave
Electromagnetic waves are called microwaves.

Mitochondria
They are eukaryotic cellular organelles responsible for supplying most of the energy necessary for cellular activity and respiration.

Neocortex
It is the name given to the most evolved areas of the cerebral cortex.

Giant Rebel Waves
They are large waves that form suddenly offshore, apparently at random. They can reach heights from peak to valley of more than 30 meters and pressures of up to 100 tons per square meter.

Oma
The suffix "oma" indicates a relationship between many molecular structures or processes at the cellular level. Ex: genome, proteome, connectome.

Planetesimals
They are celestial bodies formed within a protoplanetary disk or a disk of debris. The Planetesimal Hypothesis states that planets are formed from grains of cosmic dust that collide and adhere to form larger and larger bodies.

Recombinant
Recombinant or recombined DNA is an artificial molecule deliberately formed in vitro by joining DNA sequences from two different organisms that are generally not found together. A protein not present in a given organism and produced from recombinant DNA is called recombinant protein.

Relativity
The theory according to which physical laws change when the reference system is changed.

Resonance
Is the sound produced by the repercussion with another sound. For example, magnetic resonance imaging is the energy absorption by the atoms of a substance subjected to magnetic fields of specific frequencies.

Retrotransposons
Also called transposable items. They are a genetic component that is copied and pasted at different genomic locations (transposon) by converting RNA back to DNA through the reverse transcription process.

Singularity
It is a location in spacetime where the density and gravitational field of an astral body are predicted to become infinite by general relativity in a way that does not depend on the coordinate system.

System
It is a group of interrelated entities that form a unified whole.

Type 1A supernovae
It is a type of supernova that occurs in binary stellar systems (two stars orbiting each other) in which one of the stars is a white dwarf. If a white dwarf gradually accumulates mass from a binary companion or merges with a second white dwarf, in that case, its core will reach the ignition temperature for carbon fusion and within seconcs of the start of nuclear fusion.

Brain stem
It is an elongated structure constituted of white matter and gray matter. It is made up of the medulla oblongata, the bridge, and the midbrain.

TED
Technology, Entertainment, Design. is an American media organization that publishes talks online for free distribution under the banner "ideas worth spreading."

UNESCO
United Nations Educational, Scientific and Cultural Organization.

Universe
It is everything that exists physically, the sum of space (matter and energy) and time.

WMAP
Wilkinson Microwave Anisotropy Probe.

References

1. Aad, G. e. (2012). Combined search for the Standard Model Higgs boson in pp collisions. *Phys Rev D*, 86, 032003. doi:10.1103/PhysRevD.86.032003.

2. Abdalá, F. G. (2018). The Triassic Record of Cynodonts: Time of innovations in the Mammalian Lineage. En: L. Tanner: *The Late Triassic World. Topics in Geobiology*, 46: 407-445. New York: Springer Link.

3. Abdullah, M. (2020). Cosmological Constraints on Ωm and $\sigma 8$ from Cluster Abundances Using the GalWCat19 Optical-spectroscopic SDSS Catalog. Astrophys J 901(2):14pp. DOI: 10.3847/1538-4357/aba619.

4. Aij, R. e. (2015). Observation of Resonances Consistent with the Pentaquark States in Decays. *Phys. Rev. Lett,* 115, 072001(1) - 072001(15).

5. Aiola, S. (2020). The Atacama Cosmology Telescope: Dr4 Maps and Cosmological Parameters. *arXiv:2007.07288.*

6. Airapetian, V. G. (2017). How Hospitable Are Space Weather Affected Habitable Zones? The Role of Ion Escape. *Astrophys J Lett,* 836(1). doi:10.3847/2041-8213/836/1/L3.

7. Aktipis, A. (2015). Cancer across the tree of life: cooperation and cheating in multicellularity. *Phil. Trans. R. Soc. B*, 370, 20140219. doi:10.1098/rstb.2014.0219.

8. Aktipis, A. (2017). Cooperation and cheating as innovation: insights from cellular societies. *Philos Trans R Soc Lond B Biol Sci.* 372(1735): 20160421. DOI: 10.1098/rstb.2016.0421.

9. Alberto, J. A. (2013). La Tierra Origen y Constitución. *Rev Geo Dig,* 10(19).

10. Alcober, O. M. (2010). A new herrerasaurid (Dinosauria, Saurischia) from the Upper Triassic Ischigualasto Formation of northwestern Argentina. *ZooKeys,* 63, 55-81. doi:10.3897/zookeys.63.550.

11. Alfaro, J. S. (2018, Nov 14). An accelerating Universe without Λ in concordance with the last H0 measured value. Retrieved from arXiv Cornell University: https://*arxiv.org/pdf/1811.05828.pdf*

12. Allsopp, A. 1969. Phylogenetic relationships of the procaryote and the origin of the eukaryotic cell. *New Phytol.* 68:591-612.

13. Alriga (2020). The beginning of the accelerated expansion of the Universe: the acceleration of the scale factor. *The Web of Physics.* https://forum.lawebdefisica.com/blogs/alriga/316773.

14. American Psychiatric Association. (2014). DSM-5. *Diagnostic and Statistical Manual of Mental Disorders.* Buenos Aires: *Editorial Médica Panamericana.*

15. Andreotti, B. (2014). What is colour? Colour in relation to Qualia and consciousness. Retrieved from *ON QUALIA - academia.edu*: https://www.academia.edu/.

16. Anónimo. (n.d.). *Hammurabi code.* Luarna.

17. Arese, L.F. (2019). How the brain transitions from conscious to subliminal perception. *Neuroscience* 411:280-290 doi:10.1016/ j.neuroscience. 2019.03.047.

18. Argue, D. G. (2017). The affinities of *Homo floresiensis* based on phylogenetic analyses of cranial, dental, and postcranial character. *J Hum Evol,* 107-133. doi:10.1016 / j. jhevol.2017. 02.006.

19. Arlinghaus, K. B. (2016). Insects cannot tell us anything about subjective experience or the origin of consciousness. *PNAS,* 113(27), E3813. doi:10.1073 / pnas.1606835113.

20. Arsuaga, J. (2014). Neandertal roots: Cranial and chronological evidence from Sima de Los Huesos. *Science* 344(6190):1358-1363. DOI: 10.1126/science.1253958.

21. Asher, J. L. (2009). A whole-genome scan and fine-mapping linkage study of auditory-visual synesthesia reveal evidence of linkage to chromosomes 2q24, 5q33, 6p12, and 12p12. *Am J Hum Genet,* 84(2), 279-285. DOI: 10.1016/j.ajhg.2009. 01.012.

22. Ashtekar A. (2021). Black holes, Big Bang and Gravitational Waves: Examples of Paradigm Shifts in Fundamental Science. *PPGF/UFPA - Programa de Pós-Graduação em Física. https://youtu.be/889mFLCieBs.*

23. Association for the Study of Animal Behaviour *(ASAB).* (2016). Animal Navigation from Quantum Physics to Global Migration.

24. Atmanspacher, H. (2015, Jun 2). Quantum Approaches to Consciousness. Retrieved Oct 11, 2017, from *Stanford Encyclopedia of Philosophy*: https://plato.stanford.edu/entries/ qt-consciousness/#4.1.

25. Audi, R. (2004). *The Cambridge Dictionary of Philosophy* (Diccionario Akal de Filosofía) (1a ed.). Cambridge: Akal editores.

26. Avise, J. C. (2010). In the Light of Evolution IV: The Human Condition. Washington, DC: *The National Academies Press*.

27. Axe, D. D. (2010). The Limits of Complex Adaptation: An Analysis Based on a Simple Model of Structured Bacterial Populations. *BIO.complexity* (4), 1-10. doi:10.5048/BIO-C.2010.4.c.

28. Azevedo, F. A. (2009). Equal numbers of neuronal and nonneuronal cells make the human brain an isometrically scaled-up primate brain. *J Comp Neurol.*, 513(5), 532-541. doi:10.1002/cne.21974.

29. Baars B. J., E. D. (2012). Consciousness, biology, and quantum hypotheses. *Phys Life Rev*, 9(3), 285-294. DOI: 0.1016/j.plrev.2012.07.001.

30. Baltimore, D. (1971). Expression of Animal Virus Genomes. *Bact Rev*, 35(3):235-241.

31. Baragli, P. (2921). If horses had toes: demonstrating mirror self-recognition at group level in Equus caballus. *Anim Cogn* DOI: 10.1007/s10071-021-01502-7.

32. Barboni, M. (2017). Early formation of the Moon 4.51 billion years ago. *Scie Adv* 3(1): e1602365 doi: 10.1126/sciadv.1602365.

33. Barge, L. (2015) From Chemical Gardens to Chemobrionics. *Chem Rev* 115:8652-8703. https://doi.org/10.1021/ acs.chemrev.5b00014.

34. Barron, A. K. (2016). What insects can tell us about the origins of consciousness. *PNAS*, 113(18), 4900-4908. doi:10.1073/ pnas.1520084113.

35. Barrow, J. T. (1986). The Anthropic Cosmological Principle. New York: *Oxford University Press*.

36. Basell, L. S. (2008). Middle Stone Age (MSA) site distributions in eastern Africa and their relationship to Quaternary environmental change, refugia, and the evolution of Homo sapiens. *Quatern Scie Rev*, 27, 2484–2498.

37. Batygin, K. L. (2015). Jupiter's decisive role in the inner Solar System's early evolution. *PNAS*, 112(14), 4214–4217. doi:10.1073/pnas.1423252112.

38. Baumeister, R. L. (2018). On the Necessity of Consciousness for Sophisticated Human Action. *Front Psychol*, 9, 1925. doi:10.3389/fpsyg.2018.01925.

39. Beatty, J. T. (2005). An obligately photosynthetic bacterial anaerobe from a deep-sea hydrothermal vent. *PNAS*, 102(26), 9306–9310. doi:10.1073/pnas.0503674102.

40. Benner, S. A. (2010). Defining Life. *Astrobiology*, 10(10), 1021-1030. doi:10.1089/ast.2010.0524.

41. Berger, L. R. (2015). Homo naledi, a new species of the genus Homo from the Dinaledi Chamber, South Africa. *eLife*, 4, e09560. doi:10.7554/eLife.09560.

42. Bhargava, P. (2003). Panspermia—true or false? *Lancet*, 362(9381), 407. DOI: http://dx.doi.org/10.1016/S0140-6736 (03)14041-X

43. Bhattacharya, R. e. (2019, May). The Fate of Our Universe: How This Universe Will End. *RRJoSST*, 8(1), 14-27.

44. Bianconi, E. P. (2013). An estimation of the number of cells in the human body. *Ann Hum Biol*, 40(6), 463-471. doi:10.3109/03014460.2013.807878

45. Jerusalem Bible (2011). New edition, fully revised. Minneapolis: *Liturgical Press*.

46. Billings, L. (2015, Apr 1). Jupiter, Destroyer of Worlds, May Have Paved the Way for Earth. *Sci Am*.

47. Blakey, R. (2015). NAU Geology. The Paleogeography Maps. https://www2.nau.edu/rcb7/.

48. Boly, M. et al. (2013). Consciousness in humans and non-human animals: recent advances and future directions. Front Psychol, 4: 625 doi: 10.3389 / fpsyg.2013.00625.

49. Boly, M. (2017). Are the Neural Correlates of Consciousness in the Front or the Back of the Cerebral Cortex? Clinical and Neuroimaging Evidence. *J Neurosci*, 37(40), 9603-9613. doi:10.1523/JNEUROSCI. 3218-16.2017.

50. Bower, G.C. (2022). "Focus on First Sgr A* Results from the Event Horizon Telescope." *Astrophys J Lett*. https://iopscience. iop.org/journal/2041-205/page/Focus_on_First_Sgr_A_Results

51. Bradbury, J. (2011). Docosahexaenoic Acid (DHA): An Ancient Nutrient for the Modern Human Brain. *Nutrients*: 3(5): 529–554.

52. Brasier, M. (2015). Deep questions about the nature of early life signals: a commentary on Lister (1673) 'A description of certain stones figured like plants.' *Phil. Trans. R. Soc. A*, 373, 20140254. doi:10.1098/rsta.2014.0254.

53. Brockman, j. (2019). The Last Unknowns: Deep, Elegant, Profound Unanswered Questions About the Universe, the Mind, the Future of Civilization, and the Meaning of Life. New York: *William Morrow and Company*.

54. Brookes, J. (2010). Science is perception: what can our sense of smell tell us about ourselves and the world around us? *Philos Trans A Math Phys Eng Sci.*, 368(1924), 3491-3502. doi:10.1098/rsta.2010.0117.

55. Brusatte, S. B. (2015). The extinction of the dinosaurs. *Biol Rev*, 90(2), 628-642. doi:10.1111/brv.12128.

56. Buckman, E. (1980). The use of humor in psychotherapy. Boston University: *Dissert Abstr Int*, 41(5-B), 1715.

57. Buddingh, B. C. (2017). Artificial Cells: Synthetic Compartments with Life-like Functionality and Adaptivity. *Acc. Chem. Res.*, 50(4), 769-777. DOI: 10.1021/acs.accounts.6b00512.

58. Burbidge, E. B. (1957). Synthesis of the Elements in Stars. *Rev Mod Phys*, 29(4), 547-654.

59. Bushdid, C. M. (2014). Humans Can Discriminate More than 1 Trillion Olfactory Stimuli. *Science,* 343(6177), 1370-1372. doi:10.1126/science.1249168.

60. Butler, A. B. (2012). Hallmarks of consciousness. *Adv Exp Med Biol*, 739, 291-309. doi:10.1007/978-1-4614-1704-0_19.

61. Byars, S. E. (2010). Natural selection in a contemporary human population. *PNAS*, 107(Suppl 1), 1787-1792. doi:10.1073/pnas.0906199106.

62. Byrnes, J. K. (2018). Anomalous K-Pg–aged seafloor attributed to impact-induced mid-ocean ridge magmatism. *Sci Adv*, 4(2), eaao2994. DOI: 10.1126/sciadv.aao2994.

63. Cacace, V. (2011) Biology of Prions. Institute of Cardiology Research, University of Buenos Aires-Conicet.

64. Cady, L. B.-M. (2014). Where Do We Go from Here? Astrobiology Editorial Board Opinions. *Astrobiology*, 14(8), 629-644. doi:10.1089/ast.2014.1405

65. Caldwell, R. K. (2003). Phantom Energy: Dark Energy with $w < -1$ Causes a Cosmic Doomsday. *Phys Rev Lett*, 91(7), 071301. doi:10.1103/PhysRevLett.91.071301.

66. Callaway, E. (2013). Genetic Adam and Eve did not live too far apart in time. *Nature*. doi:10.1038/nature.2013.13478.

67. Calvo, P. (2019). Plants are intelligent, here is how. *Ann Bot*, 125(1):11-28.

68. Camacho, J.M. (2003). Humor and the creative dimension of psychotherapy. Psychodebate 6. Psychology, Culture, and Society. p. 45-58.
www.palermo.edu › publicaciones › pdf › Psico6.

69. Camargo, R. (2012). Mycorrhiza: A large union under the ground. *Rev Dig Univ*, 13(7):1-19 ISSN: 1067-60710.

70. Camprubí, E. (2019). The Emergence of Life. *Space Sci Rev*, 215:56 https://doi.org/10.1007/s11214-019-0624-8.

71. Canfield, D. E.-P.-L. (2013). Oxygen dynamics in the aftermath of the Great Oxidation of Earth's atmosphere. *PNAS*, 110(42), 16736-16741. doi:10.1073/pnas.1315570110.

72. Cardona, T. (2016). Reconstructing the Origin of Oxygenic Photosynthesis: Do Assembly and Photoactivation Recapitulate Evolution? *Front. Plant Sci.*, 7(257). doi:10.3389/fpls.2016.00257.

73. Carmona, G. E. (2009). Chemistry and Life. *Rev R Acad Cienc Exact Fís Nat (Esp)*, 103(2), 389-398.

74. Carrillo-Mora, P. (2010). Memory systems: historical review, classification, and current concepts. *Salud Mental*, 33, 85-93.

75. Carroll, S. (2008). Cosmic Variance. *Discover*. Retrieved from http://blogs.discovermagazine.com/cosmicvariance/2008/10/2 3/quantum-hyperion/#.WWRZXIQ1-Uk.

76. Carter, B. (1974). Large Number Coincidences and the Anthropic Principle in Cosmology. In I. A. Union, Confrontation of Cosmological Theories with Observational Data (pp. 291-298). *Dordrecht: Springer*. doi:10.1007/978-94-010-2220-0.

77. Carter, B. (2004). Anthropic principle in Cosmology. Contribution to Colloquium ``Cosmology: Facts and problems"., *arXiv:gr-qc/0606117*. Paris.

78. Castelle, C.J: (2018). Major New Microbial Groups Expand Diversity and Alter our Understanding of the Tree of Life. *Cell* 172(6):1181-1197.

79. Chalmers, D. (1995). Facing Up to the Problem of Consciousness. *J Conscious Stud*, 2(3), 200-219.

80. Chalmers, D. (2015). Why Isn't There More Progress in Philosophy? *Philosophy*, 90(351):3-31. www.jstor.org/stable/ 44015138. Accessed 13 Apr. 2021.

81. Chan, J. (2015). Hypotheses of Cancer Weakening and Origin. *J Cancer*, 6(5), 457-463. doi:10.7150/jca.10911.

82. Chan, E. (2019). Human origins in a southern African paleo-wetland and first migrations. *Nature*, 575:185-189.

83. Charitos, P. (2014). Inaugurating a space for dialogue between modern physics and religion. *Avance y 'Perspectiva*, 6(3).

84. Chatton, E. (1925). "Pansporella perplexa. Réflexions sur la biologie et la phylogénie des protozoaires" *Annales des Sciences Naturelles*. Zoologie et Biologie Animale. 10-VII. pp. 1-84.

85. Chen, I. (2006). The Emergence of Cells During the Origin of Life. *Science*, 314(5805):1558-1559.

86. Chen, S. K. (2013). New genes as drivers of phenotypic evolution. *Nat Rev Genet.*, 14(9), 645-660. doi:10.1038/ nrg3521.

87. Choi, S. (2020) The Atacama Cosmology Telescope: A Measurement of the Cosmic Microwave Background Power Spectra at 98 and 150 GHz. *arXiv:2007.07289.*

88. Chopra, D. (2014). Comment on Reality and consciousness: A view from the East Comment on "Consciousness in the universe: A review of the 'Orch OR' theory" by Stuart Hameroff and Roger Penrose. *Phys Life Rev,* 11, 81-82. DOI: http://dx.doi.org/10.1016/j.plrev.2013.08.002.

89. Chown, M. (2019). The letter Einstein received from Schwarzschild from the front lines of World War I helped discover black holes. BBC News World. https://www.bbc.com › mundo › noticias-49099270.

90. Cinabro, D. S. (2017). Search for Type Ia supernova NUV–optical subclasses. *Month Notices Royal Astronom Soc,* 466(1), 884-891. doi:10.1093/mnras/stw3109.

91. Clark, R. (2016, Jan 31). Clark Vision. Retrieved from Notes on the Resolution of the Human Eye: *http://www.clarkvision.com/articles/human-eye/*

92. Claverie, J.M. (2016). Giant viruses: The difficult breaking of multiple epistemological barriers. *Stud Hist Philos Biol Biomed Sci.* 59:89-99.
DOI: 10.1016/j.shpsc.2016.02.015.

93. Cleeves, L. I. (2014). The Ancient Heritage of Water Ice in the Solar System. *Science.* 345, 1590-1593 DOI: 10.1126/ science.1258055.

94. CMS Collaboration. (2021). Search for top squarks in final states with two top quarks and several light-flavor jets in proton-proton collisions at s$\sqrt{}$= 13 TeV. arXiv:2102.06976.

95. Cockell, C. S. (2003). Impossible Extinction. Natural catastrophes and the supremacy of the microbial world (1st ed.). Cambridge, United Kingdom: *University Press.*

96. Cockell, C. S. (2014). Habitable worlds with no signs of life. *Phil. Trans. R. Soc. A,* 372, 20130082. DOI :10.1098/ rsta.2013.0082.

97. Cockell, C. S. (2015). Astrobiology: Understanding Life in the Universe (1st. ed.). Singapore: *Wiley Blakey.*

98. CODATA Value (2018). Planck -time, length, mass, temperature- *NIST* (National Institute of Standards and Technology). https://www.nist.gov/pml/fundamental-physical-constants.

99. Coge, L. (2013). Implicit memory and explicit memory. Retrieved from *Kopher's blog*: https://kopher.wordpress.com/2013/09/02/la-memoria-implicita-y-la-memoria-explicita/.

100. Cohen, K. M. (2013). Chronostratigraphic Chart. *Episodes,* 36(3), 199-204.

101. Collins, G. S. (2020). A steeply inclined trajectory for the Chicxulub impact. *Nat Commun* 11(1480). https://doi.org/ 10.1038/s41467-020-15269-x.

102. Connely, J. (2016). Lead isotope evidence for a young formation age of the Earth-Moon system. *Earth Planet Sci Lett,* 452:36-43.

103. Connely, J. (2017). Pb–Pb chronometry and the early Solar System. Geochim Cosmochim Acta, 201:345-363. doi:10.1016/j.gca.2016.10.044.

104. Costanzo, A. (2017). Fat Taste Sensitivity Is Associated with Short-Term and Habitual Fat Intake. *Nutrients* 9(7):781. DOI: 10.3390/nu9070781.

105. Conselice, C. J. (2016). The Evolution of Galaxy Number Density at Z <8 and its Implications. *Astroph J,* 830(2). doi:10.3847/0004-637x/830/2/83.

106. Cornish, B. (2017). Life before LUCA *J Theor Biol.* 2017. Dec 7;434:68-74. DOI: 10.1016/j.jtbi.2017.05.023.

107. Cox, B. C. (2013). Wonders of Life: Exploring the Most Extraordinary Phenomenon in the Universe. London, *(BBC, Ed.) Harper Design.*

108. Craddock, T. H. (2017). The quantum underground is where mind and life originate. In: J. T. Poznanski, The Biophysics of Consciousness: A Foundational Approach (pp. 459-516). Singapore: *World Scientific Publishing.*

109. Craig, W. (1988). Barrow and Tipler on the Anthropic Principle vs. Divine Design. *Br J Philos Sci,* 39(3), 389-395. doi:10.1093/bjps/39.3.389.

110. Crespi, B. S. (2003). Ecology. Social slime molds meet their match. *Science*, 299(5603), 56-57.

111. Crespo, J. L. (2017). Teleportation and the Tunnel Effect. *Quantum Fracture.* *https://www.youtube.com/watch?v=wLL1VEuwmw8.*

112. Crespo, J. L. (2020). We are not made of particles. *Quantum Fracture. https://www.youtube.com/watch?v=Tsnyq-3k7Bg.*

113. Crick F.H. (1957). On Protein Synthesis. Symposium, London, Sept 1957. Retrieved from: *http://libgallery.cshl.edu/items/show/52220.*

114. Croswell, K. (1996). The Alchemy of the Heavens: Searching for Meaning in the Milky Way. Washington, DC: *Anchor.*

115. Dambeck, T. (2012) "Retuschen an der Entstehungsges chichte des Erdtrabanten". Retoques sobre la génesis de la luna de la Tierra. *Neue Zürcher Zeitung.* https://www.nzz.ch/wissen/wissenschaft/retuschen-an-der-entstehungsges chichte-des-erdtrabanten-1.17592343?reduced=true.

116. Damer, B. (2020). The Hot Spring Hypothesis for an Origin of Life. *Astrobiol* 20(4):429-452. DOI: 10.1089/ast.2019.2045.

117. Davies, P. (2003). How bio-friendly is the universe? *Int J Astrobiol,* 2(2), 115-120. doi:10.1017/S1473550403001514.

118. Davis, S. C. (2019). Undervalued potential of crassulacean acid metabolism for current and future agricultural production. *J Exp Bot* 29;70(22):6521-6537.

119. Dawkins, R. (2008). The God Delusion. New York: *Mariner Books.*

120. Deamer, D. (2008). What Is Life? What Was Life? What Will Life Be? *Prog Theor Phys Suppl,* 173, 11-16.

121. Deamer, D. W. (2017). The Role of Lipid Membranes in Life's Origin. *Life,* 7(5). doi:10.3390/life7010005.

122. Deamer, D. W. (2017). Can Life Begin on Enceladus? A Perspective from Hydrothermal Chemistry. Astrobiology, 17(9):834-839. doi:10.1089/ast.2016.1610.

123. Deamer, D. W. (2010). Bioenergetics and Life's Origins. *Cold Spring Harb Perspect Biol*, 2, a004929. doi:10.1101/ cshperspect. a004929.

124. Deamer, D. W. (2015). Hydrothermal Conditions and the Origin of Cellular Life. *Astrobiology*, 15(12). doi:10.1089/ ast.2015.1338.

125. Derrida, J. (2008). The Gift of Death. University of Chicago Press.

126. Détroit, F. M. (2019). A new species of Homo from the Late Pleistocene of the Philippines. *Nature*, 568(7751), 181-186. doi:10.1038/s41586-019-1067-9.

127. Devlin, H. (2015). This is the way the world ends: not with a bang, but with a Big Rip. *The Guardian*. https://www. theguardian.com/science/2015/jul/02/not-.with-a-bang-but-with-a-big-rip-how-the-world-will-end.

128. Dicke, R. H. (1961). Dirac's Cosmology and Mach's Principle. *Nature*, 192: 440–441. https://doi.org/10.1038/192440a0.

129. Di Giulio, M. (2011). The Last Universal Common Ancestor (LUCA) and the Ancestors of Archaea and Bacteria were Progenotes. *J Mol Evol*, 72, 119-126. doi:10.1007/s00239-010-9407-2

130. Diéguez, C. (2003). Flora and Vegetation during the Jurassic and Cretaceous. *Monograf Jard Bot Cordob*, 11, 53-62.

131. Diener, T. (2016) Viroids: "living fossils" of primordial RNAs? *Biol Dir* (2016) 11:15 DOI 10.1186/s13062-016-0116-7

132. Dinan, T. S. (2015). Collective unconscious: how gut microbes shape human behavior. *J Psychiatr Res*, 63, 1-9. DOI: 10.1016/j.jpsychires.2015.02.021.

133. Dirks, P. E. W. (2017). The age of *Homo naledi* and associated sediments in the Rising Star Cave, South Africa. *eLife*, 6, e24231. doi:10.7554/eLife.24231

134. Dodd, M. P. (2017). Evidence for early life in Earth's oldest hydrothermal vent precipitates. *Nature*, 543, 60-64. doi:10.1038/nature21377.

135. Donald, M. (1990). Quantum Theory and the Brain. *Proc Roy Soc Lond A*, 427, 43-93.

136. Drake, F. (1961). Drake Equation. Project Ozma. Green Bank Observatory. *SETI Institute.* https://www.seti.org/understanding-drake-equation.

137. Dugne, J. J. (2002). Preon Trinity - A Schematic Model of Leptons, Quarks, and Heavy Vector Bosons. *Europhys Lett,* 60(2), 188-194. doi:10.1209/epl/i2002-00334-5.

138. Durrer, R. (2011). What do we really know about Dark Energy? *Phil Trans R Soc A, 369(1957), 5102-5114.* doi:10.1098/rsta.2011.0285.

139. Dyson, F. (1985). Origins of Life. Cambridge: *Cambridge University Press.*

140. EHTC (2019). First M87 Event Horizon Telescope Results. IV. Imaging the Central Supermassive Black Hole. The Event Horizon Telescope Collaboration. *Astrophys J Lett.* 875(1):L4 (52pp).

141. EHTC (2021). First M87 Event Horizon Telescope Results. VII. Polarization of the Ring. The Event Horizon Telescope Collaboration. *Astroph J Lett.* 910: L12 (48pp).

142. Einstein, A. (1917). "Kosmologische Betrachtungen zur allgemeinen Relativitätstheorie" (Cosmological considerations on the general theory of relativity) *"Reports of meetings of the Prussian Academy of Sciences in Berlin"* 142-152.

143. Elert, G. (2018). Beyond The Standard Model. Retrieved from *The Physics Hypertextbook*: https://physics.info/beyond/.

144. Elterman, K. M. (2014). Postoperative Alterations in Taste and Smell. *Anesth Pain Med,* 4(4), e18527. doi:10.5812/ aapm.18527.

145. Eme, L. (2017). Archaea and the origin of eukaryotes. *Nat Rev Microbiol.* 2017 Nov 10 ;15(12):711-723. DOI : 10.1038/ nrmicro.2017.133.

146. Encyclopedia Britannica, Inc. (2006). Britannica Encyclopedia of World Religions. Singapore: *Merriam-Webster Inc.*

147. Engels, S. H. (2012). Night-Migratory Songbirds Possess a Magnetic Compass in Both Eyes. *PLoS One,* 7(9), e43271. DOI: 10.1371/journal.pone.0043271.

148. Enns, P. (2010). When was Genesis written, and why is it relevant? Brief historical study. Science and Faith Center. *Documentos BioLogos.* https://www.fliedner.es/media/modules/editor/cienciayfe/docs/biologos/Documento_BioLogos_Enns_Genesis.pdf.

149. Eriksson, K. R. (1991). From the Big Bang to sustainable societies. *Acta Oncol,* 30(6), 5-14.

150. Espichán, J. (2017). Introduction to Quantum Field Theory. The National University of Callao. *Technical report.* http://repositorio.unac.edu.pe/handle/UNAC/2347

151. Exoplanet TEAM. (2022). *The Extrasolar Planets Encyclopaedia.* Retrieved from http://exoplanet.eu/.

152. Fajman, D. (2020). Future Attractors in 2+1 Dimensional Λ Gravity. *Phys Rev Lett,* 125(12):121102. DOI: 10.1103/PhysRevLett.125.121102.

153. Fasching, L. (2015). TRIM28 Represses Transcription of Endogenous Retroviruses in Neural Progenitor Cells. *Cell Reports,* 10(1):20-28.

154. Feinberg, T. (2013). The evolutionary and genetic origins of consciousness in the Cambrian Period over 500 million years ago. *Front Psychol,* 4, 667. doi:10.3389/fpsyg.2013.00667.

155. Fisher, M. (2015). Quantum cognition: The possibility of processing with nuclear spins in the brain. *Ann Phys,* 362(2015), 593-602. DOI: 10.1016/j.aop.2015.08.020.

156. Forgan, D.H. (2009). A numerical testbed for hypotheses of extraterrestrial life and intelligence. Int J Astrobiol 8(2):121-131. https://doi.org/10.1017/S1473550408004321.

157. Forterre, P. (2009). The great billion-year war between ribosome- and capsid-encoding organisms (cells and viruses) as the major source of evolutionary novelties. *Ann N Y Acad Sci;*1178:65-77. DOI: 10.1111/j.1749-6632.2009.04993.x.

158. Forterre, P. (2010). Defining Life: The Virus Viewpoint. *Orig Life Evol Biosph,* 40, 151-160.doi:10.1007/s11084-010-9194-1.

159. Forterre, P. (2015). The universal tree of life: an update. *Front Microbiol.,* 6(717). doi:10.3389/fmicb.2015.00717.

160. Fox G. E. (2010). Origin and Evolution of the Ribosome. *Cold Spring Harb Perspect Biol.* 2010 Sep; 2(9): a003483. DOI: 10.1101/cshperspect.a003483.

161. Fox, R. S. (2011). A new, early Puercan (earliest Paleocene) species of Purgatorius (Plesiadapiformes, Primates) from Saskatchewan, *Canada. J Paleontol,* 85(3), 537-548. doi:10.1666/10-059.1.

162. Franco, R. (2021). What is oxidative stress, and how does it affect aging. *The Conversation.* Universitat de Barcelona.

163. Fraser, G. S. (2014). Potential solar axion signatures in X-ray observations with the XMM-Newton observatory. *Mon Not R Astron Soc,* 445(2), 2146-2168. doi:10.1093/mnras/stu1865.

164. Freud, S. (1940 / 2013). The joke and its relationship with the unconscious. Buenos Aires: V.5. Hyspamerica.

165. Freud, S. (1940 / 2013). Scheme of Psychoanalysis. Buenos Aires: Paidós.

166. Friedman, A. (1922). Über die Krümmung des Raumes» (On the Curvature of Space) *Zeitschrift für Physik (Revista de física)* 10:377ss.

167. Friedman, A. (1924). Über die Möglichkeit einer Welt mit konstanter negativer Krümmung des Raumes» (On the possibility of a World with Constant Negative Curvature in Space). *Zeitschrift für Physik (Journal of physics)* 21:326-332.

168. Frijia, E. (2021). Functional imaging of the developing brain with wearable high-density diffuse optical tomography: A new benchmark for infant neuroimaging outside the scanner environment. *NeuroImage* 225:117490. DOI: 10.1016/ j. neuroimage.2020.117490.

169. Friston, K. (2013). Life as we know it. *J R Soc Interface,* 10, 20130475. doi:10.1098/rsif.2013.0475.

170. Gambini, R. (2020). Loop Quantum Gravity for Everyone. Singapore. *World Scientific Publishing Company.* ISBN-13: 978-9811211959.

171. García-Salcedo, R. M. (2007). Description of the evolution of the Universe: a presentation for pre-university students. *Lat. Am. J. Phys. Educ.,* 1(1), 95-100.

172. Garstang, M. K. (2017). Understanding Animal Detection of Precursor Earthquake Sounds. *Animals (Basel).*, 7(9), E66. doi:10.3390/ani7090066

173. Gasperini, M. V. (2003). The Pre-Big Bang Scenario in String Cosmology. *Phys Rept*, 373, 1-212. doi:10.1016/S0370-1573(02)00389-7.

174. Gauger, E. R. (2011). Sustained Quantum Coherence and Entanglement in the Avian Compass. *Phys Rev Lett*, 106(4), 040503. doi:10.1103/PhysRevLett.106.040503

175. Gerhard, O. (2008). Pattern speeds in the Milky Way. Mem. *S.A.It*, 00, 183-187.

176. Gerrit, B. (2019). Molybdenum isotopic evidence for the late accretion of outer Solar System material to Earth. *Nat Astronom* 3:736-741.

177. Ghosh, S. S. (2014). Evidence of massive global synchronization and the consciousness Comment on "Consciousness in the universe: A review of the 'OrchOR' theory" by Hameroff and Penrose. *Phys Life Rev*, 11, 83-84.

178. Gibbons, A. (2015). Revolution in Human Evolution. *Science*, 349(6246), 362-366.

179. Gibney, E. (2014). Earth has water older than the Sun. *Nature* 345(6204):1590-3 doi:10.1038/nature.2014.16011.

180. Gibson, D. G. (2010). Creation of a bacterial cell controlled by a chemically synthesized genome. *Science*, 329(5987), 52-56. doi:10.1126/science.1190719.

181. Gigerenzer, G. (2008). Gut Feelings: The Intelligence of the Unconscious. New York: *Penguin Books*.

182. Giulio, D. (2010). Biological evidence against the panspermia theory. *J Theor Biol*, 266(4), 569-572. DOI: 10.1016/ j.jtbi.2010.07.017.

183. Glansdorff, N. X. (2008). The Last Universal Common Ancestor: emergence, constitution and genetic legacy of an elusive forerunner.*Biol Dir*, 3(29).doi:10.1186/1745-6150-3-29.

184. Glossary of philosophy. (2018). Mind. Retrieved from Glossary of Philosophy: https://www.webdianoia.com/glosario/display.php?action=view&id=211&from=action=search|by=M.

185. Gogarten, J. D. (2016). Is LUCA a thermophilic progenote? *Nat Microbiol*, 1(16229). doi:10.1038/NMICROBIOL.2016.229.

186. Goldenfeld, N. (2014). Looking in the right direction, Carl Woese and evolutionary biology. *RNA Biology*, 11(3), 248-253. doi:10.4161/rna.28640.

187. Gomes, R. (2005). Origin of the cataclysmic Late Heavy Bombardment period of the terrestrial planets. *Nature* 435:466-469.

188. Gonzalez de Posada, F. (2004). The beginning of the first principles: The Anthropic Principle. *Anales Real Acad Nal Med* (1): 23-52.

189. Gonzalez, G. R. (2004). The Privileged Planet: How Our Place in the Cosmos Is Designed for Discovery. Washington DC: *Regnery Publishing*.

190. Gonzalez-Mestres, L. (2016). Big Bang, inflation, standard Physics... and the potentialities of new Physics and alternative cosmologies. *EPJ Web of Conferences*, 126, 02012. doi:10.1051/epjconf/201612602012.

191. Gott J. R. (2005). A Map of the Universe. *Astrophys J*, 624, 463-484.

192. Gowanlock, M. G. (2011). A Model of Habitability Within the Milky Way Galaxy. *Astrobiology*, 11(9), 855-873. doi:10.1089/ast.2010.0555.

193. Grande, A. H. (2009). Environment, energy for life, and solar electromagnetic radiation. *Rev R Acad Cienc Exact Fís Nat (Esp)*, 103(2), 341-356.

194. Granqvist, P. N. (2017). Nature Meets Nurture in Religious and Spiritual Development. *Br J Dev Psychol.*, 35(1), 142-155. doi:10.1111/bjdp.12170.

195. Grant, R. H. (2011). Groundwater chemistry changes before major earthquakes and possible effects on animals. *Int J Environ Res Public Health*, 8(6), 1936-1956. doi:10.3390/ ijerph8061936.

196. Graziano, M. (2014). Speculations on the evolution of awareness. *J Cogn Neurosci*, 26(6), 1300-1304. DOI: 10.1162/jocn_a_00623.

197. Greaves, J. C. (2018, Apr 4). Paucity of phosphorus hints at a precarious path for extraterrestrial life. Retrieved from *Royal Astronomical Society*: https://phys.org/news/2018-04-paucity-phosphorus-hints-precarious-path.html

198. Greshko, M. (7 de Oct de 2019). The Discovery of 20 new moons gives Saturn a solar system record. Retrieved from *National Geographic:* https://www.nationalgeographic.com/science/2019/10/discovery-20-new-moons-gives-saturn-solar-system-record/.

199. Guevara-Erra, R. M. V. (2016). Statistical mechanics of consciousness: Maximization of the information content of the network is associated with conscious awareness. *Phys Rev E*, 94(5), 052402. DOI: https://doi.org/10.1103/PhysRevE.94.052402.

200. Guterstam, A. B. (2015). Decoding illusory self-location from activity in the human hippocampus. *Front Hum Neurosci*, 9(412). doi:10.3389/fnhum.2015.00412.

201. Guth, A. H. (1981). Inflationary universe: A possible solution to the horizon and flatness problems. *Rev Phys* 23(2):347-356.

202. Gutzwiller, M. (2008, Oct 27). Quantum Chaos. Sci Am.

203. Haladjian, H. M. (2016). Artificial consciousness and the consciousness-attention dissociation. *Conscious Cogn,* 45, 210-225. DOI: 10.1016/j.concog.2016.08.011.

204. Halberg, F. C. (2008). Cycles Tipping the Scale between Death and Survival (= "Life"). *Prog Theor Phys Suppl* (173), 153-181.

205. Hales, C. (2014). The origins of the brain's endogenous electromagnetic field and its relationship to the provision of consciousness. *J Integr Neurosci.,* 13(2), 313-361. doi:10.1142/S0219635214400056.

206. Halliday, A. N. (2000). Terrestrial accretion rates and the origin of the Moon. *Earth Planet Sci Lett,* 176(1), 17-30. doi:10.1016/S0012-821X(99)00317-9.

207. Hammeroff, S. (1998). Quantum computation in brain microtubules? The Penrose–Hameroff 'Orch OR' model of consciousness. *Phil Trans R Soc Lond A* 356, 1869–1896, 356, 1869-1896.

208. Hameroff, S. (2014). Quantum effects in the understanding of consciousness. *J Integr Neurosci,* 13(2), 229-252. DOI: 10.1142/S0219635214400093.

209. Hameroff, S. (2020). 'Orch OR' is the most complete and most easily falsifiable theory of consciousness. Cogn Neurosci 24:1-3 DOI: 10.1080/17588928.2020.1839037.

210. Hanczyc, M. (2014). Droplets - unconventional protocell model with life-like dynamics and room to grow. *Life (Basel),* 4(4), 1038-1049. doi:10.3390/life4041038.

211. Hanczyc, M. M. (2007). Mineral surface-directed membrane assembly. *Orig Life Evol Biosph,* 37(1), 67-82.

212. Hansford, S. H. (2014). Boveri at 100: Theodor Boveri and genetic predisposition to cancer. *J Pathol.,* 234(2), 142-145. doi:10.1002/path.4414.

213. Harbi, D. (2014). Classifying prion and prion-like phenomena. *Prion,* 8(2):161-165. DOI: 10.4161/pri.27960.

214. Harcourt-Smith, W. T. (2015). The foot of Homo naledi. Nat Commun, 6, 8432. doi:10.1038/ncomms9432.

215. Hartmann, W. D. (1975). Satellite-Sized Planetesimals and Lunar Origin. *Icarus,* 24, 504-515.

216. Hashizume, H. (2012). Role of Clay Minerals in Chemical Evolution and the Origins of Life. In M. M. Valaskova, Clay Minerals in Nature - Their Characterization, Modification, and Application (pp. 191-208). *Rijeka, Croatia: InTech* 2012. doi:10.5772/2708.

217. Hatch, N. (2020). Galaxy cluster illuminates the cosmic dark ages. *Nature* 577, 36-37. DOI: 10.1038/d41586-019-03893-7.

218. Hawking, S. (1988). A Brief History of Time: From the Big Bang to Black Holes. México: *Grijalbo.*

219. Hawking, S. M. (2010). The Grand Design (1a ed.). México: *Editorial Paidós Mexicana.*

220. Hawks, J. (2017). New fossil remains of Homo naledi from the Lesedi Chamber, South Africa. *eLife* 6:1-63. doi.org/10.7554/eLife.24232.

221. Hayakawa, M. (2013). Possible Electromagnetic Effects on Abnormal Animal Behavior Before an Earthquake. *Animals (Basel),* 3(1), 19-32. doi:10.3390/ani3010019.

222. Hayward, A. (2017). Origin of the retroviruses: when, where, and how? *Curr Opin Virol,* 25:23-27. DOI: 10.1016/j.coviro. 2017.06.006.

223. Hawking, S. (1970). The singularities of gravitational collapse and cosmology. *Proc Royal Soc A*; 314(1519):529-548.

224. Hazen, R. M. (2013). The Story of Earth: The first 4.5 billion years from stardust to living planet (1st ed.). New York: *Penguin Books.*

225. Hedges, S. B. (2009). The Timetree of Life (1st ed.). *OUP Oxford.*

226. Hedges, S. B. (2015). Tree of Life Reveals Clock-Like Speciation and Diversification. *Mol. Biol. Evol,* 32(4), 835-845. doi:10.1093/molbev/msv037.

227. Heidmann, T. (2009) Endogenous retroviruses: from infectious elements to bona fide genes with a physiological role. *Retrovirology,* 6(Suppl 2):I2 doi:10.1186/1742-4690-6-S2-I2.

228. Hershkovitz, I. (2021) A Middle Pleistocene Homo from Nesher Ramla, Israel. Science 372(6549):1424-1428.

229. Hess, (2021). Lightning strikes as a major facilitator of prebiotic phosphorus reduction on early Earth. *Nat Commun* 12(1):1535. DOI: 10.1038/s41467-021-21849-2.

230. Hilbig, G. (2010). Reconsidering "evidence" for fast-and-frugal heuristics. *Psychon Bull Rev.,* 17(6), 923-930. doi:10.3758/PBR.17.6.923.

231. Hildt, E. (2019). Artificial Intelligence: Does Consciousness Matter? *Front Psychol* 10:1535. doi: 10.3389/fpsyg.2019. 01535.

232. Hobson, J. H. (2014). Virtual reality and consciousness inference in dreaming. *Front Psychol,* 5(1133). doi: 10.3389/fpsyg.2014.01133.

233. Holloway, R. H. (2018). Endocast morphology of Homo naledi from the Dinaledi Chamber, South Africa. *Proc Natl Acad Sci U S A.,* 115(22), 5738-5743. doi:10.1073/pnas.1720842115.

234. Hongzhan, F. Y. (2017). A nearly water-saturated mantle transition zone is inferred from mineral viscosity. *Sci Adv,* 3(6), e1603024. doi:10.1126/sciadv.1603024.

235. Hood, B. (2014). The Domesticated Brain. London: *Penguin Books.*

236. Horning, D. J. (2016). Amplification of RNA by an RNA polymerase ribozyme. *Proc Natl Acad Sci U S A.,* 113(35), 9786-9791. doi:10.1073/pnas.1610103113.

237. Hubble, E. (1929). A relation between distance and radial velocity among extra-galactic nebulae. *PNAS,* 15, 168-173.

238. Hubble, E. (1929). A spiral nebula as a stellar system, Messier 31. *Astroph J,* 69, 103-158. DOI: 10.1086/143167.

239. Hublin, J. N. (2015). Brain ontogeny and life history in Pleistocene hominins. *Philos Trans R Soc Lond B Biol Sci.,* 370(1663). doi:10.1098/rstb.2014.0062.

240. Hug, L. (2016). A new view of the tree of life. *Nat Microbiol* 1, 16048. https://doi.org/10.1038/nmicrobiol.2016.48.

241. Hughes, J. (2019). The Quintessential Dark Energy Theory: Quintessence. *Math Phys.* A Preprint.

242. Hupé, J. D. (2015). A critical review of the neuroimaging literature on synesthesia. *Front Hum Neurosci,* 9(103), 1-37. doi:10.3389/fnhum.2015.00103.

243. Huterer, D. S. (2018). Dark energy two decades after observables, probes, consistency tests. *Rep Prog Phys,* 81(1), 016901. doi:10.1088/1361-6633/aa997e.

244. Huttenlocher, P. D. (1997). Regional differences in synaptogenesis in the human cerebral cortex. *J Comp Neurol,* 387(2), 167-178.

245. IAU (2018). International Astronomical Union. *Resolution B4*. 30th General Assembly (Oct. 26, 2018).

246. ICTV (2020). International Committee on Taxonomy of Viruses. https://talk.ictvonline.org/taxonomy/.

247. Isenman, L. (2013). Understanding unconscious intelligence and intuition: "blink" and beyond. *Perspect Biol Med.*, 56(1), 148-166. doi:10.1353/pbm.2013.0007.

248. Izhikevich, E. E. (2008). Large-scale model of mammalian thalamocortical systems. *Proc Natl Acad Sci U S A*, 105(9), 3593-3598. doi:10.1073/pnas.0712231105.

249. Jäkel, F. S. (2016). An overview of quantitative approaches in Gestalt perception. *Vision Res*, 126, 3-8. DOI: 10.1016/j.visres.2016.06.004.

250. Jedlicka, P. (2017). Revisiting the Quantum Brain Hypothesis: Toward Quantum (Neuro)biology? *Front. Mol. Neurosci*, 10(366). doi:10.3389/fnmol.2017.00366.

251. Jheeta, S. (2017). The Landscape of the Emergence of Life. *Life,* 7(27). doi:10.3390/life7020027.

252. Jiménez, C. (2013). "Viruses are Life!" (V European Congress of Virology, Lyon, September 11-14, 2013). *Fundación Conocimiento Madrid.* https://www.madrimasd.org/blogs/virusemergentes/2013/10/los-virus-son-vida-v-congreso-europeo-de-virologia-lyon-11-14-de-septiembre-de-2013/

253. Jin, L. K. (2018). Fatty Acid/Phospholipid Blended Membranes: A Potential Intermediate State in Protocellular Evolution. *Small,* 14(15), 20 pp. doi:10.1002/smll.201704077.

254. Joordens, s. W. (2013). Intelligence as It Relates to Conscious and Unconscious Memory Influences. *Can J Exp Psychol.,* 67(3), 165-174. doi:10.1037/a0033660.

255. Joyce, G. (2012). Bit by Bit: The Darwinian Basis of Life. *PLoS Biol,* 10(5), e1001323. DOI: 10.1371/journal.pbio.1001323.

256. Kafatos, M. K. (2016). The quantum universe: philosophical foundations and oriental medicine. *Integr Med Res*, 5(4), 237-246. DOI: 10.1016/j.imr.2016.08.003.

257. Kaku, M. (2011). America Has A Secret Weapon. Michio Kaku discusses H1B Visas and US Education System. http://youtu.be/NK0Y9j_CGgM.

258. Kane, G. (2003). Physics beyond the Standard Model. *Sci Am*, 288(6), 68-75.

259. Kapheim, K. P. (2015). Social evolution. Genomic signatures of evolutionary transitions from solitary to group living. *Science*, 348(6239), 1139-1143. doi:10.1126/science. Aaa 4788.

260. Karmin, M. e. (2015). A recent bottleneck of Y chromosome diversity coincides with a global culture change. *Genome Res*, 25, 459-466. doi:10.1101/gr.186684.114.

261. Katsuki, F. (2014). Bottom-up and top-down attention: different processes and overlapping neural systems. *Neuroscientist.* 20(5):509-21. DOI: 10.1177/1073858413514136.

262. Keller, G. (2011). Cretaceous-Tertiary Mass Extinction: Theories and Controversies. *SEPM Special Publication* (100), 7-22.

263. Kelley, M. (2017). Apparent ionospheric total electron content variations prior to major earthquakes due to electric fields created by tectonic stresses. *J Geophys Res Space Phys*, 122, 6689-6695. doi:10.1002/2016JA023601.

264. Key, B. A. (2016). Insects cannot tell us anything about subjective experience or the origin of consciousness. *Proc Natl Acad Sci U S A*, 113(27), E3813. doi:10.1073/pnas. 1606835113.

265. Kim, C. (2018), Artificial strain of human prions created in vitro. *Nat Commun*, 4;9(1):2166. DOI: 10.1038/s41467-018-04584-z.

266. Kindleysides, S. (2017). Fat Sensation: Fatty Acid Taste and Olfaction Sensitivity and the Link with Disinhibited Eating Behaviour. *Nutrients* 9(8):879. DOI: 10.3390/nu9080879.

267. Kipp, M.A. (2017). Biomass recycling and Earth's early phosphorus cycle. *Sci Adv.* 3(11), eaao4795. DOI: 10.1126/ sciadv. aao4795.

268. Kivell, T. D. (2015). The hand of Homo naledi. *Nat Commun*, 6, 8431. doi:10.1038/ncomms9431.

269. Klaers, J. S. (2010). Bose-Einstein condensation of photons in an optical microcavity. *Nature*, 468(7323), 545-548.

doi:10.1038/nature09567.

270. Klein, R.G. (2016). Issues in human evolution. *PNAS*, 113(23):6345-6347.

271. Knudson, A. (1971). Mutation and Cancer: Statistical Study of Retinoblastoma. *PNAS*, 68(4), 820-823. doi:10.1073/pnas. 68.4.820.

272. Koch, C. M. (2016). Neural correlates of consciousness: progress and problems. *Nat Rev Neurosci*, 17(5), 307-321. doi:10.1038/nrn.2016.22.

273. Koenig, P. S. (2008). Medial temporal lobe involvement in an implicit memory task: evidence of collaborating implicit and *explicit memory systems from FMRI and Alzheimer's disease. Cereb Cortex,* 18(12), 2831-2843. doi:10.1093/cercor/bhn043.

274. Koonin, E. (2006). The ancient Virus World and evolution of cells. *Biol Direct* 1:29.

275. Koonin, E. (2013) A virocentric perspective on the evolution of life. *Curr Opin Virol,* 3(5):546-57. DOI: 10.1016/j.coviro. 2013.06.008.

276. Korf, J. (2015). Quantum and Multidimensional Explanations in a Neurobiological Context of Mind. *The Neuroscientist,* 21(4), 345-355. doi:10.1177/1073858414542250.

277. Kramer, K. (1986). World Scriptures: An Introduction to Comparative Religions. New York/Mahwah: *Paulist Press.*

278. Kring, D. G. (2017). Chicxulub and the Exploration of Large Peak-Ring Impact Craters through Scientific Drilling. *GSA Today,* 27(10), 4-8. doi:10.1130/GSATG352A.1.

279. Kruijer, T. B. (2017). Age of Jupiter inferred from the distinct genetics and formation times of meteorites. *PNAS,* 114(26), 6712-6716. doi:10.1073/pnas.1704461114.

280. Krupovic, M. (2020). The LUCA and its complex virome. Nat Rev Microbiol; 18(11):661-670. DOI: 10.1038/s41579-020-0408-x.

281. Kunimoto, M. (2020). Searching the Entirety of Kepler Data. II. Occurrence Rate Estimates for FGK Stars. *Astronom J,* 159 (6): 248 DOI: 10.3847/1538-3881/ab88b0.

282. Landaw, J. B. (2011). Budismo para Dummies (1a. ed.). Barcelona: *Planeta*.

283. Landenmark, H. K. (2015). An Estimate of the Total DNA in the Biosphere. *PLoS Biol*, 13(6), e1002168. DOI: 10.1371/journal.pbio.1002168.

284. Lane, N. A. (2010). How did LUCA make a living? Chemiosmosis in the origin of life. *BioEssays*, 32, 271–280.

285. Lanza, M. B. (2010). Biocentrism: How Life and Consciousness are the Keys to Understanding the True Nature of the Universe (1st ed.). Dallas, TX. *BenBella Books*. doi:978-1935251743.

286. Lanza, M. B. (2017). Beyond Biocentrism: Rethinking Time, Space, Consciousness, and the Illusion of Death. Dallas, TX. *BenBella Books*. doi:978-1944648657.

287. Larrañaga, A. (2008). Big Bang Models in String Cosmology. Colombian Congress of Astronomy and Astrophysics, (pp. 54-59). Medellín, Colombia.

288. Last, C. (2014). Human Evolution, Life History Theory, and the End of Biological Reproduction. *Curr Aging Sci*, 7(1), 17-24. doi:10.2174/1874609807666140521101610.

289. Legname, G. (2017). The Prion Concept and Synthetic Prions. *Prog Mol Biol Transl Sci*. 2017;150:147-156. DOI: 10.1016/bs.pmbts.2017.06.002.

290. Lemaître, G. (1927). Un Univers homogène de masse constante et de rayon croissant rendant compte de la vitesse radiale des nébuleuses extra-galactiques. *Ann de la Soc Scient de Bruxeles*, 47, 49-59. doi:1927ASSB...47...49L.

291. Lemaître, G. (1931). The Beginning of the World from the Point of View of Quantum Theory. *Nature,* 127(706), 706-707. doi:10.1038/127706b0.

292. León, P. G.-G. (2007). The chloroplast: a key organelle in the life and use of plants. *Biotecnología,* 14, 223-238.

293. Lewalter, K. M. (2006). Bioenergetics of archaea: Ancient energy-conserving mechanisms developed in the early history of life. *BBA* (1757), 437-445.

294. Li, T. (2019). The finer scale of consciousness: quantum theory. *Ann Transl Med,* 7(20):585. DOI: 10.21037/ atm.2019.09.09.

295. LIGO. (2016, Feb 11). LIGO Opens New Window on the Universe with Observation of Gravitational Waves from Colliding Black Holes. Retrieved from *GW150914 Press Release:* https://www.ligo.caltech.edu/page/press-release-gw150914.

296. Lincoln, D. (2012). The Inner Life of Quarks. *Sci Am,* 307(5), 36-43. doi:10.1038/scientificamerican1112-36.

297. Linde, A. (1998). Universe, Life, Consciousness. Quantum Cosmology and the Nature of Consciousness, section 9: Paper delivered at the *Physics and Cosmology Group.*

298. Lingman, M. L. (2019). Relative Likelihood of Success in the Search for Primitive versus Intelligent Extraterrestrial Life. *Astrobiology,* 19(1), 28-39. doi:10.1089/ast.2018.1936.

299. Lithgow, T. S. (2010). Evolution of macromolecular import pathways in mitochondria, hydrogenosomes, and mitosomes. *Phil. Trans. R. Soc. B,* 365, 799-817. doi:10.1098/rstb. 2009.0167.

300. Livio, M. (2011, Nov 10). Lost in translation: Mystery of the missing text solved. *Nature,* 479, 171-173. doi:10.1038/ 479171a.

301. Lleó, A. (2015). Enigmas of the Universe that were clarified during the 18th century. In A. Lleó, The Great Enigmas of the Universe and the Sages in Charge of Unveiling Them. From the Dawn of Civilization to the Present (2a ed.). *Madrid: UPM Press.*

302. Loeb, L. L. (2003). Multiple mutations and cancer. *PNAS,* 100(3), 776-781.

303. Loeb, A. (2016). On the Habitability of Our Universe. *arXiv:1606.08926 [astro-ph.CO].*

304. Longhurst, C. E. (2013). Islamic Cosmogony According to Sūrah Al-Ḥashr Verse 24: "He is Allāh, the Creator, the Originator, the Fashioner of Forms." *Eur Sci J,* 2, 116-125.

305. López, C. A. (2018). Disk stars in the Milky Way detected beyond 25 kpcs from its center. *A&A,* 612, 4 pp. DOI: 10.1051/0004-6361/201832880.

306. López G.I. (2015). Retroviruses, "fossilized" viruses in our genome, which affect the functioning of the brain. Retrieved from: *microBIO*. https://microbioun.blogspot.com/2015/01/retrovirus-virus-fosilizados-en-nuestro.html.

307. López, S. (2016). Human Dispersal Out of Africa: A Lasting Debate. *Evol Bioinform Online*, 11(Suppl 2), 57-68. doi:10.4137/EBO.S33489.

308. Lorenzo, C. N. (2012). New human fossil to the last Neanderthals in central Spain (Jarama VI, Valdesotos, Guadalajara, Spain). *J Hum Evol*, 62(6), 720-725. doi: 10.1016/j.jhevol.2012.03.006.

309. Low, P. (2012). The Cambridge Declaration on Consciousness. Francis Crick Memorial Conference. Cambridge. Retrieved from http://fcmconference.org/

310. Lucas, S. L. (1993). Adelobasileus from the Upper Triassic of West Texas: The Oldest Mammal. *J Vertebr Paleontol*, 309-334. DOI: 10.1080/02724634.1993.10011512.

311. Lui, C. C. (2011). SETG: An instrument for detection of life on Mars ancestrally related to life on Earth. *Aerospace Conf*, 1-12. doi:10.1109/AERO.2011.5747299.

312. Luisi, P. (1998). About various definitions of Life. *Orig Life Evol Biosph*, 28, 613-622.

313. Luminet, J.-P. (2011). Editorial note to Georges Lemaître, The beginning of the world from the point of view of quantum theory. *Gen Relativ Gravit*, 2911-2928. DOI: DOI 10.1007/s10714-011-1213-7.

314. Ma, J. (2014). Prion disease and the 'protein-only hypothesis.' *Essays Biochem*, 56:181-91. DOI: 10.1042/bse0560181.

315. Ma, W. (2016). The essence of Life. *Biology Direct* (2016) 11:49 doi:10.1186/s13062-016-0150-5.

316. Maiese, K. (2019). Locked-in syndrome. *MSD Manual. Version for Professionals.* https://www.msdmanuals.com/es-mx/professional/trastornos-neurol%C3%B3gicos/coma-y-deterioro-de-la-conciencia/s%C3%ADndrome-de-enclaustramiento.

317. Masaoka, K. B. (2013). Number of discernible object colors is a conundrum. *J Opt Soc Am A Opt Image Sci Vis*, 30(2), 264-277. doi:10.1364/JOSAA.30.000264.

318. Mashian, N. (2016). CEMP stars: possible hosts to carbon planets in the early Universe. *Mon Not R Astron Soc.* 460(3), 2482-2491.

319. Maslow, A.H. (1992). Motivation and Personality. Madrid: *Díaz de Santos S.A.*

320. Mateos, D. W.-V. (2017). Consciousness as a global property of dynamic brain activity. *Phys Rev E.*, 96((6-1)), 062410. doi:10.1103/PhysRevE.96.062410.

321. McFadden, G. I. (2004). Evolution: Red Algal Genome Affirms a Common Origin of All Plastids. *Curr Biol,* 14, R5114-R516. DOI: 10.1016/j.cub.2004.06.041.

322. McFadden, J. (2002). Synchronous firing and its influence on the brain's electromagnetic field: Evidence for an electromagnetic field theory of consciousness. *J. Conscious.* Stud., 9(4), 23-50.

323. McKay, C. P. (2014). Requirements and limits for life in the context of exoplanets. PNAS, 111(35), 12628–12633. DOI: 10.1073/pnas.1304212111

324. McKemmish, L. R. (2009). Penrose-Hameroff orchestrated objective-reduction proposal for human consciousness is not biologically feasible. *Phys Rev E Stat Nonlin Soft Matter Phys.,* 80((2 Pt 1)), 021912. doi:10.1103/PhysRevE.80.021912.

325. McLean, E.L. (2016). Unraveling the evolution of uniquely human cognition. *PNAS*, 113(23):6348-6354.

326. McLean, P.D. (1990). The Triune Brain in Evolution: Role in Paleocerebral functions. New York: *Springer.*

327. Meibom, S: (2013). The same frequency of planets inside and outside open clusters of stars. *Nature*, 499,55-58 doi:10.1038/nature12279.

328. Méndez, F. L. (2013). An African American Paternal Lineage Adds an Extremely Ancient Root to the Human Y Chromosome Phylogenetic Tree. *AJHG* 92(3):454-459. https://doi.org/10.1016/j.ajhg.2013.02.002.

329. Menor-Salvan, C. (2013). The chemistry of the origin of life. *An. Quím.,* 109(2), 121-129.

330. Messerschmidt JL (2017). Cancer Clonal Theory, Immune Escape, and Their Evolving Roles in Cancer Multi-Agent Therapeutics. *Curr Oncol Rep,* 19(10):66. DOI: 10.1007/s11912-017-0625-2.

331. Meteoritical Bulletin Database (2022). *The Meteoritical Society. International Society of Meteoritics and Planetary Science.* Last update: 17 Apr 2022.

332. Meyer, M. (2012). A High Coverage Genome Sequence from an Archaic Denisovan Individual. *Science,* 338(6104), 222-226. doi:10.1126/science.1224344.

333. Meyer, M. (2016). Nuclear DNA sequences from the Middle Pleistocene Sima de Los Huesos hominins. *Nature* 531(7595):504-7. DOI: 10.1038/nature17405

334. Min, B. (2010). A thalamic reticular networking model of consciousness. *Theor Biol Med Model,* 7(10). doi:10.1186/ 1742-4682-7-10

335. Minot, S. (2011). The human gut virome: inter-individual variation and dynamic response to diet. *Genome Res.* 2011 Oct;21(10):1616-25. DOI: 10.1101/gr.122705.111.

336. Mocombe, P. Consciousness in the Universe According to Phenomenological Structuralism. *J Cult Soc Anthropol,*1(1):1-9.

337. Moore, A. (2008). The future of our species. *EMBO Rep, 9(Suppl 1),* S1-S3. doi:10.1038/embor.2008.111.

338. Morandi, A. S. (2016). Probing dark energy via galaxy cluster outskirts. *Mon Not R Astron Soc,* 457(3), 3266-3284. doi:10.1093/mnras/stw143.

339. Moreira, D. (2009). Ten reasons to exclude viruses from the tree of life. Nat Rev Microbiol. 2009 Apr;7(4):306-11. DOI: 10.1038/nrmicro2108.

340. Moreira, M. (2009). The Standard Model of Particle Physics. *Rev Bras Ens Fis,* 31(1), 1306-1322.

341. Moreno, T. (2018). What is Loop Quantum Gravity? Platzi/*Curiosamente* 129. https://www.youtube.com/watch?v=etTVe0B9ng&ab_channel= CuriosaMente.

342. Morgan, J. A. (1980). Chemical composition of Earth, Venus, and Mercury. *PNAS,* 77(12), 6973-6977.

343. Mormann, F. D. (2011). A category-specific response to animals in the right human amygdala. *Nat Neurosci,* 14(10), 1247-1249. doi:10.1038/nn.2899

344. Morsella, E. G. (2016). Homing in on consciousness in the nervous system: An action-based synthesis. *Behav Brain Sci,* e168. doi:10.1017/S0140525X15000643.

345. Mougari, S. (2019). Virophages of Giant Viruses: An Update at Eleven. *Viruses,* 11(8):733. DOI: 10.3390/v11080733.

346. Multiple planet systems. (2022). Retrieved from *The Extrasolar Planets Encyclopaedia*: http://exoplanet.eu/catalog/.

347. Muñoz, V. (2003). A Walk-Through Geometry - The Shape of the Universe. In R. M. Ibañez, divulgaMat (Vol. 8, pp. 121-140). Madrid: *Consejo Superior de Investigaciones Científicas.*

348. Mustafin, R. (2019). The Role of Reverse Transcriptase in the Origin of Life. Biochem (Moscow) 4(8):870-883.

349. Nacagawa, T. N. (2019). Dynamics in the Uppermost Lower Mantle: Insights into the Deep Mantle Water Cycle Based on the Numerical Modeling of Subducted Slabs and Global-Scale Mantle Dynamics. *Annu Rev Earth Planet Sci,* 47, 41-66. doi:10.1146 / annurev-earth-053018-060305.

350. *NASA/WMAP Science Team* (January 24, 2014). "Universe 101: What is the Universe Made Of?

351. *NASA Exoplanet Exploration* (2022). https://exoplanets.nasa.gov/tess/.

352. Nasir, A. (2014). The distribution and impact of viral lineages in domains of life. Front Microbiol, 30;5:194. doi: 10.3389/fmicb.2014.00194.

353. Natarajan, S. (2019). Resolution of Big Rip Singularity in Loop Quantum Cosmology and Bouncing Scenarios. *J Appl Phys Sci Int,* 11(1):30-37.

354. Nasir, A. (2014). The distribution and impact of viral lineages in domains of life. Front Microbiol, 30;5:194. doi: 10.3389/fmicb.2014.00194.

355. National Cancer Institute. (20 Oct 2019). The Cancer Genome Atlas Program. Retrieved from: *National Institutes of Health*. https://www.cancer.gov/about-nci/organization/ccg/research/structural-genomics/tcga.

356. NCBI Taxonomy. (2018, Feb 26). Homo sapiens neanderthalensis. Retrieved from *Taxonomy Browser*: https://www.ncbi.nlm.nih.gov/Taxonomy/Browser/wwwtax.cgi?name=Neanderthal.

357. Nemchin, A. (2008). A light carbon reservoir recorded in zircon-hosted diamond from the Jack Hills. *Nature* 454:92-95.

358. Nesbitt, S. B. (2013). The oldest dinosaur? A Middle Triassic dinosauriform from Tanzania. *Biol Lett*, 9(1), 20120949. doi:10.1098/rsbl.2012.0949.

359. Newell, B. S. (2014). Unconscious influences on decision making: a critical review. *Behav Brain Sci*, 37(1), 1-19. doi:10.1017/S0140525X12003214.

360. Niskama, S. (2015). Life and consciousness – The Vedāntic view. *Commun Integr Biol*, 8(5), e1085138. doi:10.1080/19420889.2015.1085138.

361. Noonan, J. (2010). Neanderthal genomics and the evolution of modern humans. *Genome Res*, 20(5), 547-553. doi:10.1101/gr.076000.108

362. Nordling, C. (1953). A New Theory on the Cancer-inducing Mechanism. *Br J Cancer*, 7(1), 68-72.

363. Nowell, P. H. (1960). A minute chromosome in human chronic granulocytic leukemia. *Science*, 132, 1497.

364. NRAO. (2016). Which Planet Orbits our Sun the Fastest? *National Radio Astronomy Observatory*. https://public.nrao.edu/ask/which-planet-orbits-our-sun-the-fastest/

365. Ohtani, E. Y. (2018). Fate of water transported into the deep mantle by slab subduction. *J Asian Earth Sci*, 167, 2-10. DOI: 10.1016/j.jseaes.2018.04.024

366. Ohtani, E. Y. (2020). The role of water in Earth's mantle. *Nat Sci Rev*, 7(1):224-232 https://doi.org/10.1093/nsr/nwz071.

367. Oparin, A. (1923/2004). The origin of life. Barcelona, España; *Ediciones Océano*.

368. Oppermann, H. L. (1979). Uninfected vertebrate cells contain a protein that is closely related to the product of the avian sarcoma virus transforming gene (src). *PNAS*, 76(4), 1804-1808.

369. O'Raifeartaigh, C. (2018). Investigating the legend of Einstein's "biggest blunder" *Phys Today* doi:10.1063/PT. 6.3.20181030a (30-Oct-2018).

370. Organ, C. N. (2011). Phylogenetic rate shifts in feeding time during the evolution of Homo. *Proc Natl Acad Sci U S A.,* 108(35), 14555-14559. doi:10.1073/pnas.1107806108.

371. Ortega del Campo, S. (2020). El Origen de los Virus. *Encuentros en Biología.* 13(173):26-30.

372. Ortinski, B. M. (2004). Neuronal Mechanisms of Conscious Awareness. *Arch Neurol,* 61(7), 1017-1020. doi:10.1001 / archneur.61.7.1017

373. Osinski, G. (2020). The Role of Meteorite Impacts in the Origin of Life. *Astrobiology* 20(9): 1121–1149.

374. Otsuka, K. C. (2008). Chronomics and "Glocal" (Combined Global and Local) Assessment of Human Life. *Prog Theor Phys Suppl* (173), 134-152.

375. Owen, J. (2009, Nov 24). Future Humans: Four Ways We May, or May Not, Evolve. Retrieved from *National Geographic News*: https://news.nationalgeographic.com/news/2009/11/091124-origin-of-species-150-darwin-human-evolution.html.

376. Paixão-Côrtes, V. V. (2012). Homo sapiens, Homo neanderthalensis and the Denisova specimen: New insights on their evolutionary histories using whole-genome comparisons. *Genet Mol Biol,* 35(4 (suppl)), 904-911.

377. Palczewska, G. V. (2014). Human infrared vision is triggered by two-photon chromophore isomerization. *Proc Natl Acad Sci U S A.,* 111(50), E5445–E5454. doi:10.1073/pnas. 1410162111.

378. ParaMicroBio, (February 13, 2019). Classification of Viruses. Scientific popularization blog in the field of biology. http://paramicrobio.blogspot.com/2018/10/clasificacion-de-los-virus.html

379. Parrilla, D. (2016). The Possibility of Quantum Neuroscience, according to Roger Penrose. *Themata Rev Filosofia* (54), 191-214. doi:10.12795/themata.2016.i54.10.

380. Patel, a. (2020). New-onset anosmia and ageusia in adult patients diagnosed with SARS-CoV-2 infection. *Clin Microbiol Infect*, 26(9): 1236–1241.

381. Pearce, B. (2018). Constraining the Time Interval for the Origin of Life on Earth. *Astrobiol* 18(3):343-364. https://doi.org/10.1089/ast.2017.1674.

382. Pearson, D. B. (2014). Hydrous mantle transition zone indicated by ringwoodite included within a diamond. *Nature*, 507(7491), 221-224. doi:10.1038/nature13080.

383. Penrose, R. (1989). The Emperor's New Mind: Concerning Computers, Minds, and the Laws of Physics. New York: *Oxford University Press*.

384. Penrose, R. (2011). Consciousness in the Universe: Neuroscience, Quantum Space-Time Geometry, and Orch OR Theory. *Journal of Cosmology*, Vol. 14.

385. Penzias, A. W. (1965). Measurement of the Flux Density of CAS an at 4080 Mc/s. Astrophys J, 1142, 1149-1155. doi:10.1086/148384.

386. Pettinari, G. (2015). The Standard Model of Cosmology. In G. Pettinari, The Intrinsic Bispectrum of the Cosmic Microwave Background (pp. 13-50). Hampshire - Rome: *Springer.* Retrieved from https://arxiv.org/pdf/1405.2280.pdf.

387. Piaget, J. (1926) *The representation of the world in the child.*

388. Piaget, J. (1945) *The formation of the symbol in the child.*

389. Pierre Auger Collaboration (2020). The Anatomy of Ultrahigh-Energy Cosmic Rays. *Physics* 13, 145. https://physics.aps.org/articles/v13/145.

390. Piccinini, G. (2010). The Mind as Neural Software? Understanding Functionalism, Computationalism, and Computational Functionalism. *Phyl Phenom Res* 81(12):269-311.

391. Pierens, A. R. (2014). Outward Migration of Jupiter and Saturn in 3:2 or 2:1 Resonance in Radiative Disks: Implications for the Grand Tack and Nice Models. *Astrophys J Lett.,* 795 (L11),(6pp). doi:10.1088/2041-8205/795/1/L11.

392. Planck Collaboration. (2014). Planck 2013 results. XVI. Cosmological parameters. *A&A,* 571((A16)), 1-66. DOI: 10.1051/0004-6361/201321591.

393. Planck Collaboration. (2016). Planck 2015 results in XIII. Cosmological parameters. *A&A,* 594(A13). doi:10.1051/0004-6361/201525830.

394. Planck Collaboration. (2018, Jul 17). Planck 2018 results. I. Overview and the cosmological legacy of Planck. Retrieved from *arXiv. Cornell University*: https://arxiv.org/pdf/1807. 06205.pdf.

395. Plaxco, K. W. (2011). Astrobiology: A Brief Introduction (2nd ed.). Baltimore: *Johns Hopkins University Press.*

396. Player, T.C: (2018). Posner qubits: spin dynamics of entangled Ca9(PO4)6 molecules and their role in neural processing. *J R Soc Interface,* 15(147): 20180494. DOI: 10.1098/rsif.2018. 0494.

397. Ponce de Leon, S. L. (2003). Panspermia—true or false? *Lancet,* 362(9381), 406-407. DOI: http://dx.doi.org/10.1016/ S0140-6736(03)14040-8.

398. Pross, A. P. (2013). The origin of life - what we know, what we can know, and what we will never know. *Open Biol,* 3, 120190. doi:10.1098/rsob.120190.

399. Radin, D. I. (2012). Consciousness and the double-slit interference pattern: Six experiments. *Phys. Essays,* 25(2), 157-171.

400. Radin, D. M. (2013). Psychophysical interactions with a double-slit interference pattern. *Phys Essays,* 26(4), 553-566. doi:10.4006/0836-1398-26.4.553.

401. Radin, D. S. (2015). Distant Healing Intention Therapies: An Overview of the Scientific Evidence. *Glob Adv Health Med,* 4((Suppl)), 67-71. DOI: 10.7453/gahmj.2015.012.suppl.

402. Ramanujan. (2013, Feb 9). The quantum enigma. overlap and collapse. Retrieved Oct 2017, 13, from Stubborn navigators in oceans of uncertainty: *http://ramanujan25449.blogspot.mx/2013/02/el-enigma-quantum-superposition-y.html?view =classic.*

403. Ramos, O. (2016, Nov 3). Wave Function Collapse: The Mental Power to Create Reality! Retrieved from YouTube: *https://www.youtube.com/watch?v=64tEQMLKTyc&t=72s.*

404. Range, F. V. (2015). Tracking the evolutionary origins of dog-human cooperation: the "Canine Cooperation Hypothesis." *Front Psychol,* 5(1582), 1-10. doi:10.3389/fpsyg.2014.01582.

405. Ratcliff, W. C. (2012). Experimental evolution of multi-cellularity. *PNAS,* 109(5), 1595–1600. doi:10.1073/pnas. 1115323109

406. Royal Spanish Academy and Association of Academies of the Spanish Language. (2014). Spanish dictionary (23a ed.). Madrid: *Espasa.*

407. Reggia J. (2013). The rise of machine consciousness: studying consciousness with computational models. *Neural Network;* 44:112-31 DOI: 10.1016/j.neunet.2013.03.011.

408. Reich, E. (2011). Edwin Hubble in translation trouble. *Nature News.* doi:10.1038/news.2011.385

409. Relethford, J. H. (2008). Genetic evidence and the modern human origins debate. *Heredity,* 100, 555–563. doi:10.1038/hdy.2008.14.

410. Renne, P. D. (2013). Time scales of critical events around the Cretaceous-Paleogene boundary. *Science,* 339(6120), 684-687. doi:10.1126/science.1230492.

411. Restrepo, C. (2015). Freud's Technique and its effects on Lacan's Clinic until 1964. *Master's Thesis,* Universidad de Antioquia, Department of Psychoanalysis, Medellín.

412. Ribas, E. D. (2012). Permanent anosmia and ageusia after resection of a left temporoinsular low-grade glioma: anatomy-functional considerations. *J Neurosurg,* 116(5), 1007-13. doi:10.3171/2012.2.JNS111982.

413. Riccomagno, M. K. (2015). Sculpting Neural Circuits by Axon and Dendrite Pruning. *Annu Rev Cell Dev Biol,* 31, 779-805. doi:10.1146/annurev-cellbio-100913-013038.

414. Richards, M. A. (2015). Triggering of the largest Deccan eruptions by the Chicxulub impact. *Geol Soc Am Bull,* 127(11-12), 1507-1520. doi:10.1130/B31167.1.

415. Riess, A. C. (2018, Feb 9). New Parallaxes of Galactic Cepheids from Spatially Scanning the Hubble Space Telescope: Implications for The Hubble Constant. Retrieved from *arXiv.org Cornell University:* https://arxiv.org/pdf/1801. 01120.pdf.

416. Riess, A. C. (2019). Large Magellanic Cloud Cepheid Standards Provide a 1% Foundation for the Determination of the Hubble Constant and Stronger Evidence for Physics beyond ΛCDM. *Astrophys J,* 876(85), 13 pp. doi:10.3847/ 1538-4357/ab1422.

417. Riser, R. O. (2017). Power Spectrum of Long Eigenlevel Sequences in Quantum Chaotic Systems. *Phys Rev Lett,* 118(20), 204101. doi:10.1103/PhysRevLett.118.204101.

418. Rizal, Y. (2020). Last appearance of Homo erectus at Ngandong, Java, 117,000-108,000 years ago. *Nature* 577(7790):381-385. DOI: 10.1038/s41586-019-1863-2.

419. Roberts, A. L. (Director). (2017). The Day the Dinosaurs Died [Motion Picture]. Retrieved from *https://www.youtube.com/ watch?v=Id-M1lerGDQ*

420. Robinson, G. B. (2017). Epigenetics and the evolution of instincts. *Science,* 356(6333), 26-27. doi:10.1126/science. aam6142.

421. Rosenberg, K. D. (2017). Evolution of the Human Pelvis. *Anat Rec (Hoboken),* 300(5), 789-797. doi:10.1002/ar.23580.

422. Rowe, T. (1988). Definition, Diagnosis, and Origin of Mammalia. *J Vertebr Paleontol,* 8(3), 261-264.

423. Rozett, R. S. (2000). Encyclopedia of the Holocaust. London, New York: Routledge, *Taylor and Francis Group.*

424. Rucker, R. (2008). Everything Is Alive. *Progr Theor Phys Suppl.* 173, 363-370. doi:10.1143/PTPS.173.363

425. Rufu, R. A. (2017). A multiple-impact origin of the Moon. *Nat Geosci,* 10, 89-94. doi:10.1038/ngeo2866.

426. Ruggiero M.A: (2015). A higher-level classification of all living organisms. *PLoS One* 10(6): e0130114. DOI: 10.1371/journal.pone.0130114. eCollection.

427. Running, C.A. (2013). Fat taste in humans: sources of within- and between-subject variability. *Prog Lipid Res.* 52(4):438-45. doi:10.1016/j.plipres.2013.04.007.

428. Sagan, C. (1985). Cosmos (8a ed.). Barcelona: *Planeta*.

429. Sagan, C. (1996). The Demon-Haunted World. Science as a Candle in the Dark. New York: *Ballantine Books*.

430. Sahu, S. G. (2014). Multi-level memory-switching properties of a single brain microtubule. *Appl Phys Lett*, 102. doi:10.1063/1.4793995.

431. Saigusa, T. Amoebae anticipate periodic events. *Phys Rev Lett.* 100(1):018101. DOI: 10.1103/PhysRevLett.100.018101.

432. https://web.archive.org/web/20090621145535/http://www.presidencia.gob.mx/prensa/ultimasnoticias/?contenido=41945.

433. Santaolalla, J. (2016, Dec 6). What is quantum mechanics? Retrieved from *Date un Voltio*: https://www.youtube.com/channel/UCns-8DssCBba7M4nu7wk7Aw.

434. Santaolalla, J. (2020, Oct 14). What was there before the Big Bang? Retrieved from *Date un Vlog*: https://www.youtube.com/watch?v=atSi4_Wzu4I&ab_ channel=DateunVlog.

435. Santaolalla, J. (2020, Dec 31). What does it mean ... to exist? Retrieved from *Date un Vlog*: https://www.youtube.com/watch?v=jE7Wc1vMBSE&t =622s&ab_channel=DateunVlog.

436. Scardia, G. (2020). What kind of hominin first left Africa? *Evol Anthropol* doi:10.1002/evan.21863.

437. Schiffer, F. (2019). The physical nature of subjective experience and its interaction with the brain. *Med Hypotheses*, 125, 57-69. DOI: 10.1016/j.mehy.2019.02.011.

438. Schlitz, M. E. (2012). Distant healing of surgical wounds: an exploratory study. *Explore (NY)*, 8(4), 223-230. DOI: 10.1016/j.explore.2012.04.004.

439. Schneider, E. H. (2014). Epigenetic Information from Ancient DNA Provides New Insights into Human Evolution. *Brain Behav Evol*, 84, 169–171. doi:10.1159/000365650.

440. Scholtyßek, C. (2017, Jul 27). Color vision in animals: From color, blind seals to tetrachromatic vision in birds. *Ophthalmologe*, 1-7. doi:10.1007/s00347-017-0543-6.

441. Schroder, K. S. (2008). Distant future of the Sun and Earth revisited. *Mon Not R Astron,* 386(1), 155-163. doi:10.1111/j.1365-2966.2008.13022.x.

442. Schrödinger, E. (2005). What is life? (U. d. Salamanca, Ed.) Salamanca: *Textos de Biofísica.*

443. *Schulte, P. A. (2010). The Chicxulub Asteroid Impact and Mass Extinction at* the Cretaceous-Paleogene Boundary. *Science,* 327, 1214-1218.

444. Schwabe, C. (2002). Genomic Potential Hypothesis of Evolution: A Concept of Biogenesis in Habitable Spaces of the Universe. *Anat Rec*, 268, 171-179. doi:10.1002/ar.10151.

445. Schwartzberg, L. (2019). Fantastic fungi. The Mushroom movie. *Moving Art. https://fantasticfungi.com.*

446. Scotese, C. (2003). PaleoMap Project. http://www.scotese. com/.

447. Seager, S. T. (2014). Exo-S: Starshade Probe-Class Exoplanet Direct Imaging Mission Concept. Astrophysics Division Science Mission Directorate NASA, Exoplanet Exploration Program Astronomy, Physics, and Space Technology Directorate Jet Propulsion Laboratory. Pasadena, Cal.: *Science and Technology Definition Team.*

448. Seager, S. (2016). Toward a List of Molecules as Potential Biosignature Gases for the Search for Life on Exoplanets and Applications to Terrestrial Biochemistry. *Astrobiol* 16(6):465-485.

449. Seager, S. (2018). The search for habitable planets with biosignature gases framed by a 'Biosignature Drake Equation.' *Int J Astrobiol* y 17 (4): 294-302. doi:10.1017/ S147355 0417000052.

450. Sefah, K. (2014). In vitro selection with artificial expanded genetic information systems. *Proc Natl Acad Sci U S A.* 111(4):1449-54. DOI: 10.1073/pnas.1311778111.

451. Sender, R. F. (2016). Revised estimates for the number of human and bacteria cells in the body. *PLOS Biol,* 14(8), e1002533. DOI: 10.1371/journal.pbio.1002533.

452. Seth, A. I. (2006). Theories and measures of consciousness: an extended framework. *Proc Natl Acad Sci U S A,* 103(28), 10799-10804.

453. Shaw, G. (2016). Earth's Early Atmosphere and Oceans, and the Origin of Life. Switzerland: Springer International Publishing. doi:10.1007/978-3-319-21972-1.

454. Sheehan, P. M. (1986). Detritus feeding as a buffer to extinction at the end of the Cretaceous. *Geology,* 14(10), 868-870.doi:10.1130/0091-7613(1986)14<868:DFAABT>2.0.CO; 2

455. Shkoporov, A. (2019). The Human Gut Virome Is Highly Diverse, Stable, and Individual Specific. *Cell Host Microbe*;26(4):527-541.e5. doi:10.1016/j.chom.2019.09.009.

456. Skowron, D. S. (2019). A three-dimensional map of the Milky Way using classical Cepheid variable stars. *Science,* 365(6452), 478-482. doi:10.1126/science.aau3181.

457. Sleep, N. (2018). Geological and Geochemical Constraints on the Origin and Evolution of Life. *Astrobiology* 18(9):1199-1219. DOI: 10.1089/ast.2017.1778.

458. Sliney, D. (2016). What is light? The visible spectrum and beyond. *Eye (Lond)*, 30(2), 222-229. doi:10.1038/eye. 2015.252.

459. Slipher, V. (1915). Spectrographic observations of Nebulae. *Popul Astron*, 23, 21-24.

460. Smith, J. (1987). β-Mg_2SiO_4: a potential host for water in the mantle? *Am Mineral,* 72, 1051-1055.

461. Smith, Y. (2018). History of the electron microscope. *News Med Life Sci.* Retrieved from: https://www.news-medical. net/life-sciences/History-of-the-Electron-Microscope-(Spanish).aspx.

462. Smolin, L. (2004). Atoms of Space and Time. *Sci Am* 290(1):66-75. doi:10.1038/scientificamerican0104-66.

463. Solís, H. L.-H. (2009). Functional neuroanatomy of memory. *Arch Neurocien (Mex)*, 14(3), 176-187.

464. Sousa, F. L. (2014). Biochemical fossils of the ancient transition from geoenergetics to bioenergetics in prokaryotic one-carbon compound metabolism. *Biochim Biophys Acta* (1837), 964–981. DOI: 10.1016/j.bbabio.2014.02.001.

465. Springer, M. S. (2011). The historical biogeography of Mammalia. *Phil. Trans. R. Soc. B*, 366, 2478-2502. DOI: 10.1098/rstb.2011.0023.

466. Stehelin, D. (1976). DNA related to the transforming gene(s) of avian sarcoma viruses is present in normal avian DNA. *Nature* 260, 170–173.

467. Steinhardt, P. J. (2011, Abril). The Inflation Debate. *Sci Am*, 36-43.

468. Stephens, P. G. (2011). Massive Genomic Rearrangement Acquired in a Single Catastrophic Event during Cancer Development. *Cell*, 144(1), 27-40. DOI: 10.1016/j.cell.2010.11.055.

469. Stepka, Z. (2022). Hidden signatures of early fire at Evron Quarry (1.0 to 0.8 Mya). *PNAS:*119(25): e2123439119. doi: 10.1073/pnas.2123439119.

470. Stevenson, D. H. (2014). The origin of the Moon. *Phil. Trans. R. Soc. A*, 372, 20140289. doi:10.1098/rsta.2014.0289.

471. Stringer, C. (2016). The origin and evolution of Homo sapiens. *Philos Trans R Soc Lond B Biol Sci*, 371(1698), pii: 20150237. doi:10.1098/rstb.2015.0237.

472. Struyve, W. (2017). Loop quantum cosmology and singularities. *Sci Rep* 7: 8161. DOI: 10.1038/s41598-017-06616-y.

473. Symonds, N. (1986). What is Life: Schroedinger's influence on Biology. *Quart Rev Biol*, 61(2).

474. Szostak, J. (2016). On the Origin of Life. *Medicine*, 76(4), 199-203.

475. Takeuchi, N. H. (2017). Conceptualizing the origin of life in terms of evolution. *Philos Trans A Math Phys Eng Sci*, 375(2109), pii: 20160346. doi:10.1098/rsta.2016.0346.

476. Tanabashi, M. (2018) Review of Particle Physics. Phys Rev D98, 030001. *DOI: 10.1103/PhysRevD.98.030001.*

477. Tang, R. D. (2014). Biophoton signal transmission and processing in the brain. *J Photochem Photobiol B,* 139, 71-75. doi:10.1016 / j.jphotobiol.2013.12.008.

478. Tanner, J. (2019). Obliquity of the Ecliptic, Nutation in Obliquity and Latitudes of the Arctic/Antarctic Circles. *PHP Science Labs.* Retrieved from: *http://www. neoprogrammics.com/obliquity_of_ the_ecliptic/.*

479. Tarlacı, S. P. (2016). Quantum neurophysics: From non-living matter to quantum neurobiology and psychopathology. *Int J Psychophysiol.,* 103, 161-173. DOI: 10.1016/j.ijpsycho. 2015.02.016.

480. Thompson, D. R. (2014). Psychopathy: clinical features, developmental basis, and therapeutic challenges. *J Clin Pharm Ther,* 39(5), 485-495. doi:10.1111/jcpt.12182.

481. Tintino, G. (2014). From Darwinian to Technological Evolution: Forgetting the Human Lottery. *Cuad Bioet,* 25(3), 387-395.

482. TIrard, S. M. (2010). The Definition of Life: A Brief History of an Elusive Scientific Endeavor. *Astrobiology,* 10(10), 1003-1009. doi:10.1089/ast.2010.0535.

483. Tomson, S. A. (2011). The genetics of colored sequence synesthesia: Suggestive evidence of linkage to 16q and genetic heterogeneity for the condition. *Behav Brain Res,* 223(1), 48-52. DOI: 10.1016/j.bbr.2011.03.071

484. Tonello, L. C. (2015). On the possible quantum role of serotonin in consciousness. *J Integr Neurosci.,* 14(3), 295-308. doi:10.1142/S021963521550017X.

485. Tonello, L. C. (2017). Stream of consciousness: Quantum and biochemical assumptions regarding psychopathology. *Med Hypotheses,* 101, 78-84. DOI: 10.1016/j.mehy.2017.02.013.

486. Torres F.S. (2016). A proposal of the proteome before the last universal common ancestor (LUCA). *Int J Astrobiol.* 15 (Sp 1):27-31.

487. Torruella, G. (2015). Phylogenomics Reveals Convergent Evolution of Lifestyles in Close Relatives of Animals and Fungi. *Curr Biol,* 25(18), 2404-2410.

488. Trewavas, A. (2011). The ubiquity of consciousness. *EMBO Rep.* 12(12): 1221–1225. DOI: 10.1038/embor.2011.218.

489. Trewavas, A. (2016). Intelligence, Cognition, and Language of Green Plants. *Front Psychol,* 7, 588. doi:10.3389/fpsyg. 2016.00588.

490. Trifonov EN (2011). A vocabulary of Definitions of Life Suggests a Definition. *J Biomol Struct Dyn,* 29(2): 259–266.

491. Trifonov, E.N. (2016). Acytota - associated kingdom of neglected life. *J Biomol Struct Dyn.* 34(8):1641-8. DOI: 10.1080/07391102.2015.1086959

492. Trixler, F. (2013). Quantum Tunnelling to the Origin and Evolution of Life. *Curr Org Chem,* 17(16), 1758-1770.

493. Tsokolov, S. (2010). A Theory of Circular Organization and Negative Feedback: Defining Life in a Cybernetic Context. *Astrobiology,* 10(10), 1031-1042. doi:10.1089/ast.2010.0532.

494. Tsokolov, S. A. (2008). Constructing an Interdisciplinary Context for Definition of Life. *Prog Theor Phys Suppl,* 173, 351-362.

495. Turin, L. (1996). A spectroscopic mechanism for primary olfactory reception. *Chem Senses,* 21(6), 773-791.

496. Twitchett, R. (2007). The Lilliput effect in the aftermath of the end-Permian extinction event. *PALAEO,* 252(1-2), 132-144. DOI: 10.1016/j.palaeo.2006.11.038.

497. Underdown, S. O. (2016). Do patterns of covariation between human pelvis shape, stature, and head size alleviate the obstetric dilemma? *Proc Natl Acad Sci U S A.,* 113(3), E258. doi:10.1073/pnas.1517524113.

498. Uno, K.T. (2016). Neogene biomarker record of vegetation change in eastern Africa. *PNAS* 113(23):6355-6363.

499. Vaillant, G. (2000). Adaptive mental mechanisms: Their role in positive psychology. *Am Psyc,* 55(1):89-98.

500. Vakoch, D. A: (2015) The Drake Equation. Cambridge University Press.

501. Van den Bergh, S. (2011). The Curious Case of Lemaitre's Equation No. 24. *arXiv:*1106.1195 [physics.hist-ph].

502. Van Kranendonk, M. (2017). Life Springs. *Sci Am.* Jul 16; 317(2):28-35. doi: 10.1038/scientificamerican0817-28.

503. Van Seters, J. (1975). Abraham in History and Tradition. New Haven: Yale University Press.

504. Varela, F. (1981). Autonomy and autopoiesis. In G. &. Roth, Self-organizing systems: An interdisciplinary approach. (pp. 14-24). Frankfurt/New York: *Campus, Verlag.*

505. Varela, F. H. (1974). Autopoiesis: the organization of living systems, their characterization, and a model. *Curr Med Biol,* 5(4), 187-196.

506. Veeroja, K. (2020) Models for Eukaryogenesis. The hydrogen hypothesis. *Bachelor's Thesis. Faculty of Medicine and Health Technology University of Tampere.* Finland.

507. Vida, K. K. (2017). Frequent Flaring in the TRAPPIST-1 System Unsuited for Life? *Astrophys J,* 841(2). doi:10.3847/1538-4357/aa6f05.

508. Villarreal, L. P. (2011). Viral Ancestors of Antiviral Systems. *Viruses,* 3, 1933-1958. doi:10.3390/v3101933.

509. Vink, J.S. (2020). Star Formation in the Ultraviolet. *Galaxies* 8(2):43. doi.org/10.3390/galaxies8020043.

510. Vithoulkas, G. (2014). Conscience and consciousness: a definition. *J Med Life* 15;7(1):104-8.

511. Wang, Z. W. (2015). An integrated phylogenomic approach toward pinpointing the origin of mitochondria. *Scie Rep,* 5, 7949. doi:10.1038/srep07949.

512. Ward, J. (2013). Synesthesia. *Annu Rev Psychol,* 64, 49-75. DOI :10.1146/annurev-psych-113011-143840.

513. Ward, P. B. (2003). Rare Earth. Why Complex Life is Uncommon in the Universe. New York: *Copernicus Books.*

514. Ward, P. B. (2004). The Life and Death of Planet Earth: How the New Science of Astrobiology Charts the Ultimate Fate of Our World. New York: *Holt Paperbacks.*

515. Watkins, L. v. (2019). Evidence for an Intermediate-mass Milky Way from Gaia DR2 Halo Globular Cluster Motions. *The Astrophysical Journal,* 873(2), 118. doi:0.3847/1538-4357/ab089f.

516. Watson, M. A. (2014). Synesthesia and learning: a critical review and novel theory. *Front Hum Neurosci*, 8(98), 1-15. doi:10.3389/fnhum.2014.00098

517. Watson, M. R. (2017). The prevalence of synaesthesia depends on early language learning. *Conscious Cogn*; 48:212-231. doi:10.1016/j.concog.2016.12.004.

518. Webb, B. (2012). Cognition in insects. *Philos Trans R Soc Lond B Biol Sci*. 367(1603):2715-2722. doi: 10.1098/rstb.2012.0218.

519. Weingarten, C. D. (2016). A New Spin on Neural Processing - Quantum Cognition. *Front Hum Neurosci,* 10(541). doi: 10.3389/fnhum.2016.00541.

520. Wells, C. (2019). Prionoid Proteins in the Pathogenesis of Neurodegenerative Diseases. *Front Mol Neurosci* 12(271) doi: 10.3389/fnmol.2019.00271.

521. Whiten, A. E. (2012). The human socio-cognitive niche and its evolutionary origins. *Phil. Trans. R. Soc. B,* 367, 2119–2129. doi:10.1098/rstb.2012.0114.

522. Wieczorek, R., 2012. On prebiotic ecology, supramolecular selection and autopoiesis. *Orig. Life Evol. Biosph.*, 42, 445–450. doi:10.1007/s11084-012-9306-1.

523. Wilczec, F. (2015). A weighty mass difference. *Nature*, 520, 303-304. doi:10.1038/nature14381.

524. Will, C. (2014). Short-range modifications of Newtonian gravity. In: The Confrontation between General Relativity and Experiment *Living Rev Relativity* 17(4):27.

525. Willerslev, E. H. (2003). Panspermia—true or false? *Lancet,* 362(9381), 406. DOI : http://dx.doi.org/10.1016/S0140-6736 (03) 14039-1.

526. Woese, C. (1987). Bacterial evolution. *Microbiol Rev*, 51(2), 221-271.

527. Wong, K. (2012). The Origin of the Genus Homo. *Invest. y Ciencia* (429), 16-25.

528. Xie, S. Mycelium Book. Project Proposal. Mycelia, Fungi, and Neural Networks. https://biotechdesign.artscinow.com/node/1556.

529. Yahweh Elohim. (2015), The Torah: Messianic Hebrew Study Bible. South Carolina: *Create Space Independent Publishing Platform.*

530. Yamaguchi, M. (2012). «Prokaryote or eukaryote? A unique microorganism from the deep sea.». *J Electron Microsc (Tokyo)* 61 (6): 423-431. doi:10.1093/jmicro/dfs062.

531. Yamaguchi, M. (2014). Deep-sea microorganisms and the origin of the eukaryotic cell. *Jpn. J. Protozool.* 47(1, 2):29-48.

532. Yamaguchi, M. (2018). Electron Microscopy and Structome Analysis of Unique Amorphous Bacteria from the Deep Sea in Japan. *Cytologia* 83(3): 337–342.

533. Yamaguchi, M. (2020). Deep-Sea Bacteria Harboring Bacterial Endosymbionts in a Cytoplasm?: 3D Electron Microscopy by Serial Ultrathin Sectioning of Freeze-Substituted Specimen. *Cytologia* 85(3): 209–211.

534. Yankovic, N. B. (2010). Biology: Remarkable Experiments to Refute the Theory of Spontaneous Generation. (U. d. Talca/Chile, Ed.) Retrieved March 26, 2017, from *Portal Educativo*: http://www.educativo.utalca.cl/medios/educativo/estudiantes/media/recursos/generacion_experimentos.pdf.

535. Yang, Z. (2010). Expanded genetic alphabets in the polymerase chain reaction. *Angew Chem Int Ed Engl.* 49(1):177-80. doi:10.1002/anie.200905173.

536. Yankovic, B. (2014). Cognition, metacognition and intelligence: basic definitions. Retrieved from *http://www.educativo.utalca.cl/medios/educativo/profesores/basica/cognicion.pdf.*

537. Yitzhak, N. R. (2014). Numerical simulation of pressure waves in the cochlea induced by a microwave pulse. *Bioelectromagnetics,* 35(7), 491-496. doi: 10.1002/bem.21869.

538. Young, E. (2016). Oxygen isotopic evidence for vigorous mixing during the Moon-forming giant impact. *Science* 352(6272):493-496.

539. Yuen Chan, J. C. (2015). Hypotheses of Cancer Weakening and Origin. *J Cancer,* 6(5), 457-463. doi:10.7150/jca.10911.

540. Yufik, Y.M. (2016). Life and Understanding: The Origins of "Understanding" in Self-Organizing Nervous Systems. *Front Syst Neurosci.* 10:98 doi: 10.3389/fnsys.2016.00098.

541. Zahnle, K. S. (2010). Earth's Earliest Atmospheres. *Cold Spring Harb Perspect Biol.*, 2(10), a004895. doi:10.1101/cshperspect. a004895.

542. Zamzuri, I. (2020). Quantum Physics Perspective on Electromagnetic and Quantum Fields Inside the Brain. *Malays J Med Sci.* 2020 Feb; 27(1): 1–5. doi: 10.21315/mjms2020. 27.1.1.

543. Zhuravlev, Y.N. (2006). The definition of life in the context of its origin. *Biogeosciences* 3:281-291. www.biogeosciences. net/3/281/2006/.

544. Zinger, L. (2012). Two decades of describing the unseen majority of aquatic microbial diversity. *Mol Ecol,* 21(8):1878-1896. doi: 10.1111/j.1365-294X.2011.05362. x.

545. Zwiebach, B. (2009). A First Course in String Theory (2nd ed.). Cambridge: *Cambridge University Press.*

NOTES

FROM THE BIG BANG TO CONSCIOUS INTELLIGENCE

FROM THE BIG BANG TO CONSCIOUS INTELLIGENCE

FROM THE BIG BANG TO CONSCIOUS INTELLIGENCE

FROM THE BIG BANG TO CONSCIOUS INTELLIGENCE

FROM THE BIG BANG TO CONSCIOUS INTELLIGENCE

Author contact: **galehnycushouse@gmail.com**